LOTUS
HERITAGE

LOTUS HERITAGE

Ian Adcock

OSPREY
AUTOMOTIVE

First published in Great Britain in 1995
by Osprey, an imprint of Reed Consumer
Books Limited, Michelin House,
81 Fulham Road, London SW3 6RB and
Auckland, Melbourne, Singapore and Toronto.

ISBN 1 85532 508 X

Project Editor Shaun Barrington
Editor Julia North
Page design Paul Kime/Ward Peacock
Partnership

Printed in Hong Kong

Recommended further reading

Lotus Carlton: Ian Adcock
Illustrated Lotus Buyer's Guide: Graham Arnold
Lotus Elan Gold Portfolio 1962-1974: Brooklands Books
Lotus Europa Gold Portfolio 1966-1975: Brooklands Books
Lotus Elite & Éclat 1974-1982: Brooklands Books
Colin Chapman The Man and his Cars: Gerard Crombac
Lotus: The Elite, Elan, Europa: Chris Harvey
Colin Chapman: Lotus Engineering: Hugh Haskell
Lotus Elan: Mark Hughes
Lotus Elite: Dennis E. Ortenburger
Lotus All The Cars: Anthony Pritchard
Colin Chapman's Lotus: Robin Read
Lotus Since The '70s: Graham Robson
The Third Generation Lotuses: Graham Robson
Supercars Lotus Esprit Turbo: John Simister
Lotus Esprit: Jeremy Walton
With grateful thanks to: Lotus, Monitor, Focalpoint, the London Lotus
Centre, Mike & Gill Bishop

Half title page
*The smart new wheels of the Lotus
Esprit Turbo of 1993. See page 128.
(I. Adcock)*

Title page
*This pearlescent white Esprit, with a rear
wing for the first time, was a special
edition for the USA and was the
inspiration for a UK special.
See pages 97/98*

Right
*British Racing Green was a theme of
Lotus' advertising shortly after the second
incarnation Elan's launch. Here the
complete Lotus range is pictured with a
Lotus F1 car outside Team Lotus HQ.
(Monitor)*

Contents

Left
The Lotus Carlton (see page 120). Despite that bluff front with the wider wheel arches and the imposing front dam, the car's Cd factor still came out at a creditable 0.308. (Paul Debois)

Introduction – Lotus Road Cars

In 1986 it appeared to all the world that Lotus had fallen on its feet. After years of financial uncertainty following the untimely death of its founder, Colin Chapman, four years earlier, little Lotus was bought by the American giant, General Motors.

The asking price for the car division and Lotus Engineering was a meagre – in GMs' terms – £23 million. It seemed that, at last, Lotus had found a sanctuary in which to grow and mature, in the same manner that Fiat owned Ferrari but didn't interfere in the day-to-day running of the Prancing Horse.

During the time I was writing this book Lotus' future was, once again, full of doubt. General Motors recorded one of the biggest ever corporate losses in American history, $4.5 billion in 1992, and the universally-lauded Elan was scrapped two-and-half years and 3857 cars after it was launched.

Less than 12 months later and there was talk of a management buy-out which would have plunged Lotus back into the perilous waters from which GM had rescued it.

The MBO never happened. Managing director Adrian Palmer blamed the grey-suited bankers for not having the imagination to invest in his vision. But you can hardly blame them – the history books are littered with the corpses of failed car-makers and Lotus has had more than its fair share of close shaves.

Fortunately, a shining knight appeared in the unlikely shape of Romano Artioli. Artioli heads the secretive Bugatti International operation which was created to re-launch the fabled French marque with a range of extravagantly expensive supercars. His links with Lotus – he is the Italian importer for the marque – made him, perhaps, the logical choice as the new owner.

Of all the world's sports car manufacturers, Lotus is the most capricious. Its products can, on the one hand, be brilliant examples of ingenious automotive engineering marred, on the other, by indifferent build quality and questionable durability. This book traces thirty-six years of Lotus road cars, excluding competition variants and formula cars, and won't concern itself with Lotus' history, on or off the track, as that has

The man who started it all: Anthony Colin Bruce Chapman CBE, FRSA 1928 – 1982 .There can be no higher praise than that offered by Enzo Ferrari in Gerard Crombac's biography where he described his great rival as "a subtle visionary…and so talented because of his ability to produce ideas ahead of their time". (Focalpoint)

been well documented elsewhere. From the first Lotus production car, the Type 14 Elite, through to the current Type 80 Lotus Esprit Turbo, all have possessed a special quality and definite Lotus characteristics.

Romano Artioli and the Bugatti management is certainly imbued with the same enthusiasm, foresight and imagination that Antony Colin Bruce Chapman had when he formed the Lotus Engineering Company on 1 January, 1952. I just hope those emotions are built on an equally strong financial and business base.

Elite to Elan S3

Harold 'You've never had it so good' Macmillan was Prime Minister, Stirling Moss had finished runner-up to Juan Manuel Fangio in the Driver's World Championship and Sputnik I had been launched. The year was 1957 and Colin Chapman was about to set the motoring world alight with his Type 14. Better known as the Elite, this car lifted Lotus out of the realms of being a racing car and kit car producer into a manufacturer of road cars. And, in typical Chapman fashion, the Elite broke all the contemporary rules governing design and construction.

A conventional pressed steel body would have been prohibitively expensive and even the then traditional Lotus system of steel tubing clothed in alloy panels used for building racing cars, would have been too labour intensive and costly. However, glassfibre was becoming fashionable and, although he had loathed chemistry at school, Chapman taught himself all there was to know about this new technology.

His ability to grasp the fundamentals of grp construction, plus his engineering brilliance, resulted in a unique monocoque body which comprised of only three major components: the body, which incorporated the roof, wings and boot; a chassis comprising the floor pan, wheel wells and metal sub-frames front and rear for the suspension and final drive respectively; and what must have been an awesomely complicated single-piece structure for the engine bay and transmission tunnel, a boot moulding and mounting points for the windscreen hoop and jacking points. There were separate mouldings for the closing panels.

Equally stunning was the car's design. Penned by financier, Lotus owner and spare-time stylist, Peter Kirwan-Taylor the Elite's slippery shape is an all-time classic. Final aerodynamic tuning was performed by Frank Costin and resulted in a 0.29 Cd, a figure many cars still aspire to nearly 40 years later.

Mechanically the car was a little gem. Chapman persuaded Coventry

The original Elite which began Lotus' transformation from competition car constructor to production car manufacturer. A combined effort with styling by Peter Kirwan-Taylor, aerodynamics by Frank Costin and John Frayling who built the original clay model. (Monitor)

Climax to produce a hybrid version of its FWA racing engine; called the FWE it used the FWB's block and bore, but the FWA's shorter stroke resulting in a 1216cc capacity and 75bhp. Transmission was by an MGA-type gearbox, but with an alloy casing and a modified bell housing.

If the car was a mite underpowered its almost perfect weight distribution allied to a sophisticated suspension set up derived from the Type 12 Formula 2 car – double wishbones and coil spring/damper units at the front and Chapman strut suspension at the back* – ensured its handling was peerless. The Elite also benefited from a competition-style brake system with 9.5 inch Girling discs all round, those at the back being inboard.

It's not surprising that the specification reads more like a competition car than a road car as Chapman fully intended the Elite to be a class winner at Le Mans and in the Monte Carlo Rally.

For such events suppressing Noise, Vibration and Harshness were a waste of money and added weight so it is hardly surprising that contemporary road tests of the Elite, while praising its handling, criticised its interior noise levels: *Autocar*, 27 May, 1960; "When... various body resonances, particularly on a noise-inducing surface, coincide with the... engine period, normal conversation between driver and passenger is difficult."

The prototype Elite – *sans* radiator and prop-shaft – caused a sensation at the 1957 Earls Court motor show and orders quickly outstripped Lotus ability to build the cars. Consequently it wasn't until spring of the following year that an Elite ran in a private test session at Brands Hatch. Despite the fact that the Elite was conceived as a road car it was, for the next three years, seen primarily as a sports racer and it wasn't until 1960 – a year after Lotus moved from its Hornsey base to Cheshunt, Hertfordshire – that *Motor* magazine got to test one. By then some 280 had been built with the body/chassis made by Sussex boatbuilders, Maximar Mouldings.

Bristol Aeroplane Plastics took over the body/chassis production in July 1960 and continued building the Elite until its demise in late 1963. For 1960 the Elite received a number of small, but significant improvements; the Series II featuring longer and larger diameter front and rear springs and a new rear suspension with a triangulated strut replacing the radius arm.

At that year's Earls Court Show the Series II SE was displayed, identifiable by its factory-painted silver roof. It had an extra 10bhp under

Right
The Elite remained in production for six years until 1963 and during that time enjoyed considerable track success in the hands of drivers like Jim Clark, Innes Ireland and Cliff Allison. Unfortunately its revolutionary glass fibre monocoque construction was expensive and meant Lotus lost money on every Elite built. (Monitor)

Below right
The Lotus Elan was the car which truly established Lotus as a manufacturer and featured for the first time Chapman's rigid backbone chassis with 'Y' forks at either end to act as mounts for the suspension, transmission and engine. Lotus has stayed true to this design philosophy and over the years has steadily modified the concept still employing it today in the latest Esprit S4

* Basically another Chapman weight-saving system in which the rear suspension comprises of a single forward-facing radius rod, a long spring and damper unit and the driven half shaft which also provides lateral location of the rear wheels. It first appeared on the Lotus 12, Formula Two car.

the bonnet and an all-synchromesh ZF S4-12 gearbox with a separate aluminium bell housing.

Chapman always maintained that, even at its high selling price of £2118, Lotus lost about £100 on every Elite built. To try and boost sales Chapman reverted to his old kit-car philosophy and, from autumn 1961, the Elite was sold for home assembly saving the enthusiast £899.

Inspirational though the Elite was, Chapman knew that if Lotus was ever to become established as a serious sports car producer it would have to crack the overseas markets and, in particular, California. And that meant an open car.

Initially, Chapman wanted designer Ron Hickman – who later went on to find fame and fortune as the creator of the Black & Decker Workmate – to follow the same principals laid down for the Elite: a one-piece glassfibre moulded chassis and body shell. The trouble was that Hickman found it virtually impossible to achieve the required torsional rigidity without a roof.

To solve the problem Chapman designed a very simple backbone type structure which was cheap, simple and light to build from 16-gauge mild

The Lotus Cortina came about because of Chapman's friendship with Walter Hayes who, when editor of the Sunday Dispatch, *had employed Chapman as the paper's motoring correspondent. Following his move to Ford to revitalise its PR image, Hayes suggested to Chapman that Lotus should build a 1000 hot Cortinas, powered by Harry Mundy's twin cam version of Ford's Kent engine, for homologation into Group 2 racing. (Ford)*

steel. The 75 lbs chassis might look like Origami engineering, but the simplicity belies its 4500 lbs ft/degree torsional rigidity, although *Motor* magazine reckoned the figure to be nearer 6000.

The chassis itself comprised a 25 ins long rectangular steel box, measuring about 10.5x6ins, which formed the rigid backbone and acted as the prop shaft housing. At the front this divides into two arms running either side of the powertrain assembly to a cross member which joins the front suspension mounting points. To the rear there is a shorter 'Y' fork which encloses the final drive and provides mounting points for the rear springs and wishbones.

Suspension followed typical Lotus practice with unequal length pressed steel double wishbones and coil spring/damper units at the front, while that at the rear was a combination of Lotus Elite and Grand Prix technology – a modified Chapman Strut if you like – with a wide-based lower wishbone and built-in coil spring/damper units. In this instance the half-shafts didn't play any role in locating the rear axle, their movement being taken up by a pair of rubber doughnut-type flexible joints.

Steering was taken care of by Triumph Vitesse-type rack-and-pinion with a Herald adjustable steering column. Girling disc brakes were used all round, the 10ins ones at the rear mounted inboard of the hub carriers and the front 9.5ins discs mounted conventionally.

Determined to have a twin cam engine, Chapman went through a number of concepts before arriving at a solution: the first was a Ford

As well as Mundy's engine, the Lotus Cortina featured uprated front suspension, close ratio gearbox and heavily revised rear suspension with an A-bracket and twin tracking arms, while aluminium closing panels kept weight down. (Ford)

Consul block fitted with a 'Raymond Mays' cylinder head, but this was abandoned when Chapman heard about Ford's new one-litre short-stroke power unit. To extract more capacity and power from the engine Chapman contracted *Autocar's* technical editor, Harry Mundy – who had conceived the Facellia twin-cam while at Coventry Climax – to design the new engine.

After increasing the capacity first to 1340cc and then 1477cc, the project was abandoned when Ford revealed its new five main bearing 1.5-litre engine. Mundy then adapted his cylinder head design to fit the new engine which went on to gain race track honours in the Lotus 23 and, most famously, in the Lotus Cortina.

The DOHC's were driven by a single roller chain, while the distributor and oil pump were powered off the existing side-mounted camshaft. Both twin Weber and Dell'Orto twin-choke carburettors were used, while export models had a pair of constant vacuum Zenith-Strombergs – which were also fitted to UK specification models between November '68 and August '69. The original 1498cc engine produced 100 bhp at 5500rpm, but from the 23rd car onwards capacity went up to 1558cc and power to 105bhp at 5500rpm. Special Equipment models gained a further 10bhp and 500rpm while the yet-to-come Sprint had 126bhp at 6500rpm under the bonnet.

As a matter of interest the cylinder heads were originally cast by J. A. Prestwich of JAP engine fame until the company was bought by Villiers at which time the tooling passed on to Lotus for them to manufacture in-house. The differential and gearbox were standard Ford items.

Hickman had a tough job designing the Elite's successor, but the challenge obviously bought the best out in him for the Elan has established itself as one of the all-time classics, and was even aped 27 years later when Mazda unveiled its MX-5 Miata. It is neat, compact and uncluttered from whichever angle it's viewed, although somewhat marred when the headlights pop up. The body itself comprised of two main sections – the floor pan and wheelarches and the main body – which are bonded together and then mounted saddle-like over the backbone chassis and secured into place via 14 bobbins and bolts. Although the body is double skinned and acts as a stressed structure it wasn't intended to increase the chassis' inherent strength although, of course, it must do so in some minor way.

By autumn 1962 the Elan was complete and ready for its Earls Court motor show debut where it was an instant hit. Its immediate popularity no doubt due to an attractive price, £1495 ready to roll or £1095 in kit form, and breathtaking performance: 0-60mph in 8.7 secs and 115mph top speed if the rev limiter was ignored.

With Jim Clark rapidly becoming a household name, and only a year

The Lotus Cortina enjoyed tremendous competition success throughout the UK, Europe and the States, winning the British Saloon Car Championship on more than one occasion, the European Touring Car Challenge and the Acropolis and RAC rallies. (Ford)

away from his first Formula One World Championship, Lotus' reputation and standing within the motor industry was at an all time high and, thanks to an ex-Fleet Street editor, was on the verge of even greater achievements and glory. Walter Hayes was editor of the *Sunday Despatch* when he first met Chapman, asking him to pen the occasional motoring column for the paper. It wasn't long, though, before Hayes was head-hunted by Ford of Britain and charged with livening up the company's image so that it would appeal to the increasingly affluent, younger market.

At about the same time the Cortina range of family saloons had been launched and Ford was keen to give it a sportier image. Remembering his talented friend from his days on the *Despatch*, Hayes contacted Chapman and suggested that Lotus should develop a Group 2 competition version and build the 1000 cars necessary for homologation. Chapman didn't need asking twice.

Essentially the Lotus Cortina – Cortina Lotus to Ford – was the two-door saloon with the 1558cc 105 Bhp Elan engine under the bonnet. Changes to the front suspension remained minimal with new spring and

damper rates and a lower ride height, the rear was subjected to a more thorough going over which included locating the axle with twin tracking arms and an A-bracket and using vertical coil spring/damper units. Weight was reduced by incorporating alloy closing panels as well as alloy castings for the clutch housing, differential and remote control extensions.

Green side flashes against a white background, quarter bumpers and Lotus badging modestly claimed that this was a genuine 106mph, 0-60 mph in 9.9 secs hot saloon. Inside there were front bucket seats, remote gearchange and an alloy wood-rimmed steering wheel. Early cars suffered from loose differentials which leaked oil over the rear suspension, but this was cured in July 1964 when standard Ford items and the Cortina GT transmission, with new second gear ratios, were used; at the same time a new two-piece prop-shaft replaced the original one-piece.

The Series 3 version of the Elan represented a major update and offered a hard top for the first time. A close-ratio four-speed gearbox was later offered as an option

Throughout its three year production life, the Lotus Cortina was built at Lotus' Cheshunt factory and received the same model updates as the factory cars so, in autumn '64 'Aeroflow' ventilation became standard and in the summer of the following year the rear suspension was radically altered with semi-elliptic leaf springs and twin radius arms replacing the original set up. Self-adjusting rear brakes and Corsair 2000E gear ratios came along later the same year as well as a special equipment, 109bhp, engine.

All told 1894 Lotus Cortinas were built before the model was replaced in 1967 by the Mark II version. However, all 4032 examples of this model were built at Ford's Dagenham plant and, as such, it never received a Lotus type number.

Meanwhile development of the Elan continued apace with the Series 2 introduced in November, 1964 which incorporated larger front brake calipers and minor cosmetic improvements, including a full-width veneer fascia. In September of the following year the Series 3 range was debuted including a fixed head version; a close-ratio five-speed 'box became optional two months later and in January, 1966 the Special Equipment model debuted. With its 115bhp engine, close-ratio transmission, servo-assisted brakes and centre-lock wheels the SE represented a substantial improvement over other models in the range. A convertible version of the S3 didn't appear until June 1966.

If this wasn't enough Chapman was engrossed in designing and developing a successor for the Lotus 7. When that replacement appeared in 1966 it couldn't have been more different from the basic, front- engined, open two-seater which, in many ways, was the essence of Chapman's philosophy of simplicity and light weight.

The Europa Saga and Elan Plus 2

The mid-Sixties was a time of feverish activity for Lotus. It was established as a credible sports car manufacturer and Lotus' increasing success on race tracks around the world meant the order books for production cars were getting ever fuller. This growing respectability – from kit cars to production cars – resulted in a new class of owners, ones who did not necessarily want to be associated with the old-fashioned Lotus 7.

It was against this background that Chapman decided there should be a replacement for the venerable 7, preferably a mid-engined car that reflected and embodied Lotus' on-track engineering ingenuity. The first stumbling block was to find an appropriate powertrain as at that time the twin-cam wasn't really suitable. Moreover, Chapman wanted to distance himself from Ford, not because he was unhappy with the relationship, far from it, but to ensure that Lotus wasn't reliant on one major supplier.

Chapman found his answer at the 1964 Motor Show where Renault debuted its new 16, a five-door hatchback with front-wheel drive. Unlike most current front-wheel drive cars with their east-west engines, the Renault's power unit was mounted north-south with its rear hard against the car's bulkhead and its gearbox forwards. It was only a matter of moments before Chapman realised that turning the engine through 180 degrees would make it an ideal power unit for the embryonic Europa.

His first step in procuring engines was to contact his longtime friend and French-based motoring journalist, Jabby Crombac. In early 1965, Crombac arranged a meeting in France and Chapman concluded a deal with Renault to supply Lotus with 500 Renault 16 powertrains for the new Europa.

Although the alloy engine's bore and stroke remained unchanged at 76 x 81mm to give a 1470cc capacity, compression ratio was increased

The Europa was originally an export-only car destined for the Continent. Note the high flying buttresses at the back which made rearward visibility strictly limited. This is, in fact, the original road test car. (Monitor)

from 8.5 to 10.25:1 and the inlet valves enlarged from 1.38ins to 1.48ins, the exhaust valves remained standard. The standard single-choke Solex carburettor was replaced by a 26mm twin-choke unit mounted on a new inlet manifold. These changes resulted in a power increase from 58.5bhp at 5000rpm to 78bhp at 6000rpm, although torque dropped from 78 lbs ft at 2800rpm to 76 lbs ft at 4000rpm.

The standard Renault 7.9ins diameter clutch was retained as were the gearbox ratios, although final drive went up to 3.56 from 3.77:1. Because the engine was reversed, a new offset pinion was needed so the Europa didn't have four reverse and one forward gears. A quill shaft took drive from the clutch above the differential to the gearbox input and back to the pinion located below. Drive shafts had simple Hooke joints at both ends replacing the Renault's standard cv joints.

As 1966 and the Europa's development progressed, the car changed

Above
Lotus' first attempt at a mid-engined production sports car was the Europa. Launched in 1966, it was powered by Renault's 1.5-litre engine and matching transmission as used in the French manufacturer's 16. The deal between Lotus and the Regie was brokered by Chapman's long-time friend, Gerard Crombac. (Monitor)

Right
An early production scene of Europa's undergoing final inspection at Lotus' then new factory at Hethel. (Monitor)

from Chapman's original concept of a simple, mid-engined replacement for the 7 to a more sophisticated car in keeping with Lotus' newfound image. It wasn't a particularly easy programme, especially as Chapman felt that development director, Ron Hickman, over-complicated the car.

An Elan-type chassis was conceived using 16 swg sheet steel. Resembling a capital 'Y', the engine was located between the forks with the Elan-like front suspension − coil springs, dampers and double wishbones − mounted on a cross piece at the other end.

A boxed cross-member running above the gearbox carried a new rear suspension design incorporating long fabricated box-section radius arms pivoted from the sides of the chassis. The rear hub posts, with extensions below the wheel rim, carried the lower spring and damper mounting as well as the outer end of long transverse links to form a wide-based lower wishbone on each side.

Above
The unmistakable profile of the Elan Plus 2; developed from the original Elan, but now with seating for two children in the back it represented Chapman's first tentative moves away from kit-cars to upmarket mainstream products. (Monitor)

Left
Lotus' desire to be perceived as an upmarket sports car manufacturer, rather than a supplier of kit-cars was reflected in the publicity shots taken by its long-serving agency, Monitor, as in this image

Braking was by 9.75ins discs at the front and 8x1.5ins drums at the back, with Elan-type 13ins x 4.5J steel wheels shod with Dunlop SP41 155 tyres. John Frayling, who had worked on the Elan coupe and convertible, was responsible for designing the Europa's distinctive body with its flying buttress rear sections which completely obliterated rear three-quarter vision. However, with a an all-up weight of 13 cwts and a 0.29 Cd it's hardly surprising that Lotus claimed a 115mph top speed and zero to 60mph in about 10secs.

However, it wasn't only the Europa's exterior design that was controversial: cockpit space was decidedly restricted despite being wider than the Elan and, moreover, the seats were fixed so that swapping from short to tall drivers meant adjusting the pedals with a spanner, though a telescopic steering column made life a bit easier. Not only that, but the side windows were fixed which would have been fine if the ventilation system could cope, but its inadequacy made the cabin unbearably stuffy. And, finally, luggage space was meagre. Add all these up and it's hardly a recipe for a successful car, but what really put people off the Europa, initially, was its high price. This was nearly £1100 - not far off the Elan and some £450 more than the original target price and there were high insurance ratings too. The glass-fibre bodyshell was bonded directly to the chassis for extra rigidity which caused immense repair problems if the car was involved in a crash. Initially the little coupe was christened 'Elfin' and then 'Concorde' before 'Europe' was arrived at, but as the first batch of cars was destined only for France and other European countries the 'e' was dropped in favour of an 'a' and 'Europa' stuck.

Once that first batch of Renault engines had been used, the French company pulled out of the deal as it was coming under pressure from Alpine-Renault's Jean Redele not to supply a rival sports car manufacturer with power plants at such an advantageous price.

However, before the first 500 engines had been used, Lotus introduced the much-improved Series 2 model in 1967. The car hadn't changed mechanically, but the body was no longer bonded directly to the chassis – which helped reduce NVH as well as cut repair costs – electric windows fitted, adjustable seats installed and to improve the interior, door trimming, a veneered facia and more sound-deadening material added. Production, though, was still restricted to left-hand drive European-only models and it would be a further two years before a UK version became available.

Part of this delay could be attributed to the long-awaited introduction of the 'Metier II', better known as the Elan +2. As far back as 1963 there had been plans to develop a 2+2 Elan based on a lengthened, widened and strengthened version of the backbone chassis, but its development took a back seat to the forthcoming Coupe.

Elan Plus 2's lined up against the perimeter fence at Hethel awaiting delivery

The interior with its leather trim and extensive use of veneer was meant to reflect further Lotus' market aspirations, although the car still relied on the twin cam engine. (Monitor)

Chapman saw the +2 as essential to Lotus' future development. Many Elan owners were now like him – married with children – but didn't want a family saloon and Chapman didn't want to lose their custom to rival manufacturers.

The first running 2+2 prototype – 'Metier II' – took to the roads in 1965, but was further delayed while the car was restyled and enlarged: wheelbase went up 12ins to 96ins to provide increased passenger room, front and rear tracks were widened by seven and six inches respectively, to 54ins at the front and 55ins at the rear. To balance these increases the car's overall length increased 24ins to 169ins. Inevitably weight went up, by 336lbs and, though the car is recognisably a development of the Elan, its longer nose, broad sloping 'B' posts and sleeker tail made it, to all intents and purposes, a new car. Production was then delayed yet again, until Lotus was fully settled into its new Hethel factory.

Underneath the sleek new skin, the +2 simply continued the Elan theme with wishbone front suspension and at the back, lower wishbones with Chapman struts. Servo-assisted 10ins Girling disc brakes were fitted all round and the Lotus knock-off wheels came with Firestone F100 tyres as standard; Dunlop SP41s or Goodyear G800s could be specified at extra cost. Power came from 118Bhp version of the Lotus Cortina engine linked to a Ford Corsair gearbox, but with a lower 3.7:1 rear axle ratio, sufficient for a 120mph top speed and 0-60mph in 8.2secs.

The +2's standard equipment reflected Chapman's desire to push Lotus further upmarket: a Burr-Walnut facia, electric windows, air horns, Pye radio and even a burglar alarm were fitted (though at the expense of carpeting), while the seats were covered in leathercloth. At £1672, the +2 in kit form was £373 more costly than the Elan coupe, while factory-built +2s cost £1923. The increased cost, according to *Autosport*'s John Bolster was well worth it: "On the road, the Elan Plus 2 at once impresses... The longer wheelbase seems to make no difference to the

Above
The Special's interior was a big improvement too, with veneer fascia and, most importantly, the space to accommodate drivers taller than Colin Chapman. (Monitor)

Left
A Europa Special destined for the UK market – it can be distinguished by its low-set headlights and the optional alloy wheels and Cavallino tyres. (Monitor)

high cornering power... The car is noticeably quieter than previous Lotus models and... performance is remarkably good."

One thing the Elan wasn't was a 2+2 occasional adults, as the rear seats – separated by the wide backbone chassis – would barely accommodate two 10-year-old children. Yet, as we shall come to see the Elan+2 was a significant step in Lotus' development to becoming a manufacturer of Grand Tourers and supercars rather than a glorified kit-car producer made good.

Minor cosmetic improvements and a serious effort to improve build quality resulted in the +2 becoming the +2S in 1968 but, more significantly, it was only sold fully-built, making it the first Lotus not to be offered in component form.

The Europa eventually arrived in the UK market with right-hand drive in July the following year in the same form as it had been sold in Europe

Installing the Lotus-Ford Big Valve twin cam engine increased the Europa's top speed to 123mph after accelerating to 60mph in about seven seconds. (Monitor)

By 1971 the Plus 2S had become the Plus 2S 130, as it was now powered by the ubiquitous Big Valve engine; a year later the name changed, again, to Plus 2S 130/5 when a five-speed gearbox became optional

since its launch. Continuing the theme established by the Elan +2S the Europa S2 was only available fully-built and, at a cost of £1666, was £180 more expensive than the Elan S/E in kit form. Twelve months later, Federal-bodied Europas started being shipped to the States.

In 1969 a young engineer who had been working on Jaguar's abortive XJ13 Le Mans project joined Lotus. His first project was to engineer a Ford twin-cam version of the Europa due in 1971; that engineer's name was Mike Kimberley.

Ever since 1966, Lotus had campaigned a Group 4 racing version of the Europa that was powered by a dry-sumped Lotus-Cosworth 1.6-litre engine. Known as the Type 47 it had limited success in hands of John Miles, but the programme was discontinued in 1969 when the experimental Type 62 took over. It was the 47 which whetted the appetites of the British motoring press and led to the development of

the much quicker Type 74, better known as the Europa Twin Cam.

Outwardly the most obvious difference between the 74 and its predecessor was the absence of the flying rear buttresses, which immediately improved rearward visibility, balanced by a small front bib spoiler which reduced lift from 66lbs to 34lbs at 100mph. In fact, these improvements were the most obvious of a myriad of changes that the car underwent.

A one-inch increase in wheelbase and the front track, plus a lower floor, meant that not only was there a bigger pedal box but also sufficient leg and headroom for someone measuring 6ft 5ins, the same height as Kimberley. Other interior improvements included reshaped seats, a twist-fly-off type handbrake and a veneered facia. There was also considerable improvement to the heating and ventilating system as well as NVH.

Hardly anything was left untouched on the Europa; more powerful headlights were installed, servo-assisted brakes standardised, cable runs

Above
The Series 4 Elan came along in March 1968 in both fixed head and convertible versions; it can be distinguished by its flared arches and bonnet bulge. (Monitor)

Left
Probably the most evocative Elan of them all, the S4 Sprint in Gold Leaf Team Lotus colours and powered by the Tony Rudd-designed 126 Bhp 'Big Valve' engine. (Monitor)

improved, the gear linkage reworked in an effort to make it less recalcitrant and twin fuel tanks with a combined 12.5 gals capacity installed forward of the rear wheels.

The improvements, however, were not merely skin deep: the chassis received stronger bracing so that its torsional rigidity went up to 200 lbs/ft per degree. The suspension was subtly changed with the front and rear roll centres lowered to 3.25ins to increase the effective swing arm length which, in turn, helped reduce camber changes when the optional low profile 70 series tyres were fitted. Softer springs were fitted to improve the ride and the steering rack realigned to overcome bump steer.

The heart of the new car was, of course, the original Elan power unit producing 105bhp at 6000rpm, a 27 per cent increase over the Renault unit, and 103 lbs/ft at 4500rpm; twin Dellorto DHLA 40s replaced the original Weber 40 DCOE set up. The standard Renault transaxle was retained, forcing Lotus (in the interests of reliability) to use the 105bhp engine instead of the more powerful Sprint version, mated to a Lotus-designed bell housing which incorporated the alternator mounting and the lower rear suspension bushes.

When it came to testing the car, *Motor* achieved 60mph in 8.2secs and lapped the MIRA test track at 116.5mph, although 120mph was recorded over the flying 1/4-mile, while *Autosport*'s John Bolster considered the Twin Cam "a far better and more practical car than its predecessor... The performance is most satisfactory and one can easily out-accelerate the opposition."

Such improvements don't come cheaply and the Europa Twin Cam now cost customers £1595 in component form, or a further £400 if you wanted a factory-built car. The Europa had come a long way from Chapman's original vision of an inexpensive replacement for the 7.

Despite the feverish activity surrounding the Europa and Elan+2 the original Elan was continually being updated and improved. After the major changes which appeared in the S3, the Series 4 arrived in March 1968 with flared wheel arches to accommodate low profile 185x13 Dunlop SP Sport tyres on 4.5J x 13 rims, rocker facia switches and a +2-like bonnet bulge. But bigger things were in store for both the Elan and Elan+2.

Amidst all of this engineering development, Chapman threw himself into floating Lotus on the London Stock Exchange. He planned this move with Fred Bushell – who had been with him since his earliest days at Hornsey and was now in charge of finance – in the spring of 1968 and by October of that year shares in Group Lotus were being sold. Chapman retained 52 per cent of these which, on paper, were worth £4.5 million. The ownership of Team Lotus remained with the Chapman family.

During the late Sixties, America was beginning to clamp down on exhaust emissions and demand improved fuel consumption which inevitably lead to power losses. As the USA was an expanding export market for Lotus, Chapman was keen that the +2S and the Elan retained their competitiveness against rival marques such as Jaguar and Porsche. That meant improving the twin-cam's power and efficiency, a job which fell to Tony Rudd, Lotus' new chief engineer. Rudd, BRM's former chief engineer, and Chapman had been long time friends and rivals on the race-track, both of them acting as consultants for each other's companies. Chapman, for instance, helped BRM overcome suspension problems and Rudd helped develop the Lotus Cortina's twin-cam, and so it was in 1969 that Rudd joined Lotus.

His main task was to get Lotus' new two-litre engine into production but he also had to update the Elan's power unit. Rather than increasing its capacity Rudd redesigned the cylinder head to accept larger inlet (1.6ins), and exhaust (1.32ins) valves; the compression ratio was upped to 10.3:1 and Dellorto DHLA40 twin-choke carburettors fitted. New 'Big Valve' cam covers were designed to add stiffness and reduce oil seepage.

Taken as a whole these changes increased the engine's output to 126bhp at 6500rpm and 113 lbs/ft torque at 5500rpm. Both the Elan and +2 received the new engine in early 1971 and were renamed the Elan Sprint and Elan+2S 130.

To excite customers the Elan Sprint came finished in a number of two-tone paint finishes the most popular being the red, white and gold of Gold Leaf Team Lotus while the new +2 was distinguished from its predecessors by a silver roof. To cope with the power increase both the differential and drive shafts were strengthened and stiffer Rotoflex 'doughnut' joints stiffened in an effort to reduce surge. A new, quieter exhaust system was also installed.

Needless to say, prices went up as well as power and performance: the coupe Elan S4 Sprint now cost £1,663 and the convertible, £1,686 ready to be built. The bigger +2 at £2,626 was now heading towards Jaguar and Porsche in pricing terms.

In general the press were complimentary about the cars, although *Autocar* failed to get anywhere near Lotus' claimed performance figure of 6.2secs to 60mph, managing only eight seconds on their first attempt and seven after the test car had returned from a factory tune-up. Nevertheless, they concluded that it was "just about the fastest way of getting from A to B other than by motorcycle... Never was a sports car more a sports car than this one."

Motor was equally impressed with the +2S 130's 7.7secs to 60mph and 121mph top speed; "...this is an outstanding car for sheer driving pleasure and, together with the Elan Sprint, the best Lotus yet." Despite the Elan Sprint and +2S 130 being lauded as the best Lotuses to date, both they and the Europa were criticised for having only four-speed gearboxes when five-speeds were becoming more common, even on everyday saloons. *Motor*, for instance, wrote of the Plus 2S 130 "this Lotus, although ideally geared for acceleration and for speed-limited countries, might be less at home maintaining a cruising speed of 110mph on a continental motorway since this represents no less than 6200rpm." And John Bolster in *Autosport* remarked that "for a good many years, there has been criticism of Lotus production cars because they were too low geared for long-distance continental touring... their high engine revs, when cruising at near the maximum for hours at a time, were tiring for the occupants and caused unnecessary wear and tear."

Such black marks were eradicated in 1972 when five-speed boxes were made available throughout the Lotus range. Those for the front-engined models were based on gear clusters from Austin installed in a Lotus-manufactured gearbox; top speed was achieved in direct fourth (17.9mph per 1000rpm), while fifth gave a lazier 22.4mph per 1000rpm. Only a few Elans were built with this box, but the +2 was clumsily re-designated the +2S 130/5.

The availability of a reasonably priced five-speed transmission from Renault – as used in its 16TS model, but with Lotus' own fifth gear – meant that the Europa could benefit from the Big Valve engine's additional power.

Re-christened the Europa Special and finished in black and gold, in deference to the F1 Team's John Player Special sponsorship, this final version of the Europa would sprint to 60mph from standstill in 6.6secs and on to a 123mph maximum. While *Motor* praised the car for its "Good performance; fabulous handling and roadholding; excellent ride; relaxed at high speed" it was less complementary about other aspects of the car, "Restricted rear view; awkward pedals; inefficient heater; some poor detail finish."

By now the entire Lotus range was beginning to show its age and Chapman, Rudd and Kimberley had embarked on an ambitious programme to radically update what Lotus had to offer.

This audacious product plan centred around a new 16-valve engine being developed by Rudd which, in four-cylinder and V8 form, would power a new family of cars including a 2+2, a four-seater and a mid-engined car.

Chapman's dream of establishing Lotus as a prestige car manufacturer was further spurred on by the introduction of VAT in 1973 when the UK joined the EEC. Effectively, this put an end to kit cars and paved the way for a new generation of Lotus products.

Lotus' Own Powerplant

Before Lotus could embark on an ambitious new model programme it needed a new engine, a fact realised as far back as 1964, just two years after the twin-cam Lotus-Ford engine was launched.

A further 24 months passed before it was decided that the new Lotus powerplant would be a 150bhp, two-litre, four-valves per cylinder design developed under the guidance of Steve Sanville, head of Lotus' powertrain development. Lotus' decision to go ahead and produce its own engine – thereby freeing itself of its dependence on Ford – coincided with Coventry Climax withdrawal from motor racing allowing Chapman to quickly snap up one of their leading engineers, Ron Burr, who had worked on the four-valve-per-cylinder Coventry Climax FWMV racing engine and the aborted flat-16 F1 engine.

Lotus had given some thought to producing a BRM-designed 24-valve V6, but this was discarded when it became apparent that a 120 degree engine would have been too wide for the traditional Lotus chassis and a 60 degree design too high for the low profile bonnets Chapman envisaged. Lotus settled on a two-litre, 24-valve slant-four – mounted at 45 degrees – which could be doubled up to a 4.2-litre V8 for possible use at Indianapolis and as a road-going engine. As the engine was intended for both road and competition, it embodied the latest technology; aluminium block and head with removable wet cylinder liners, twin belt-driven overhead camshafts, an oversquare (95.25mm x 69.85mm) bore and stroke and a target output of 150Bhp.

At the 1967 Earls Court Motor Show, Vauxhall unveiled its new slant-four engine, whose bore centres were, remarkably, exactly the same as those proposed by Lotus. Chapman immediately negotiated a deal with Vauxhall – ironically owned by GM, who would later buy Group Lotus – to buy some of their cast-iron blocks so that development of Lotus' own aluminium cylinder head could be speeded up.

When first conceived, the new slant four was supposed to be the basis of an entire family of Lotus competition and road engines: Type 904: 2-litre iron block race engine. Type 905: 2-litre iron block road engine. Type 906: 2-litre sand-cast alloy block race engine. Type 907: 2-litre die-cast alloy block road engine. Type 908: 4-litre V8 alloy block

One that never made it. Lotus helped Chrysler – later Talbot – to develop a rally homologation special based on the Sunbeam three-door with the new 2.2-litre engine crammed under the bonnet. So successful was this car that the brilliant young Finnish rally driver, Henri Toivonen, won the 1980 RAC Rally in one and the following year the Talbot factory team won the World Rally Championship for Makes. Shades of Lotus Cortina development all over again. This is just a publicity shot of an Essex-painted Sunbeam-Lotus. (Focalpoint)

race engine. Type 909: 4-litre V8 alloy block road engine.

By 1968 the first of these hybrid Type 904 engines – designated by the marketing department as the LV220 for Lotus/Vauxhall 220bhp – fitted with Tecalemit-Jackson mechanical fuel injection was up and running in the Lotus 62 sports-racing cars.

A year later, now with Tony Rudd responsible for putting the engine into production, prototype 147bhp Type 905s were put through reliability trials in a Vauxhall Viva GT and a Bedford CF van. This was quickly followed by the Type 906 which, effectively, put an end to the Lotus/Vauxhall hybrids.

The development programme went through so smoothly that by 1970 the engine was completed and ready for a car. Lotus invested £550,000 in machine tooling for the new engine and desperately needed to recoup that money, but with its new model programme still someway in the distance, and the Europa and Elan too small for the new power unit, it would be nearly two years before the Type 907 engine appeared in a production car and, even then, it wouldn't be a Lotus. When the BMC and Leyland empires merged under the leadership of Sir Donald Stokes there were casualties, one of the most famous being the big Austin-Healey 3000 sports car. Determined to continue the Healey heritage Donald Healey and his son, Geoffrey, proposed to Californian entrepreneur and former Healey dealer, Kjell Qvale that they should team up and produce a new Healey sports car.

By 1968 the project was well underway and Hugo Pole penned a two-seater, later modified by Bill Towns, based round Vauxhall Viva GT running gear. Two years later Qvale bought the insolvent Jensen Motors, made Donald Healey its chairman and told Jensen's chief engineer, Kevin Beattie, to get the Jensen-Healey into production. One of the first snags they ran into was the Viva engine's lack of suitability caused by the increasingly stringent US exhaust regulations which would have sapped the Vauxhall engine of any useful power.

Despite the obvious Lotus/Vauxhall link, it was Chapman who made the first approach to Qvale offering 60 engines a week, well short of Jensen's 200 cars a week production target. Undaunted at having been rejected, Chapman approached Jensen a second time when it became obvious that Lotus' new GT wasn't nearly ready. So it was that in October, 1971 Jensen announced it would be taking up to 15,000 (!) engines a year from Lotus.

In Jensen trim, the 907 sported twin Dellortos – US cars had horizontal Zenith Strombergs – and produced 140Bhp. The engines weren't without their problems due to oil collecting around the valve gear so it didn't drain back to the oil pump quickly enough. The Lotus Elite eventually went into production in Spring 1974, powered by a

Above
All 2.2 litre Lotus engines are assembled on site at Hethel

Left
Throughout its life, the Excel's engine remained carburetted. There was much talk of fuel-injection, but it was felt that too much development time would have to be spent on then updating the chassis and interior to warrant the expenditure. Anyway, a 2+2 version of the Elan was envisaged to take over from the Excel. (I Adcock, Car courtesy of London Lotus Centre)

155bhp version of the 907, the extra power coming from larger, 38mm choke, Dellorto carburetters. Those cars destined for the USA had 140bhp engines with a lower compression ratio – 8.4 against 9.5:1 – and twin Strombergs mounted on a water-heated manifold.

Two years of infield service resulted in a number of changes to the 907 in an effort to improve reliability, although *Autocar* still thought the four-cylinder was "rough at top end". *Motor* attributed the 907's early problems to development work which was carried out on sand-cast prototypes rather than die-cast production units. Amongst the difficulties Jensen owners experienced with their Lotus power-plants were distorted cylinder liners and pistons which had a tendency to 'pick up', while Lotus improved engine cooling by 10 degrees centigrade by fitting a skeleton-type toothed belt guide at the front of the engine.

The mid-seventies were turbulent times, not only in the UK with strikes and a three-day week, but also in the Middle East where the after-effects of the Arab-Israeli war of 1973 would have dire effects on the world's motor industry. Suddenly petrol was being restricted to the Western nations by Arabian countries, who realised they had a means of holding the West to ransom. (A mite simplified, granted.) Petrol prices soared; Jensen collapsed in 1976. With it went Lotus' engine contract.

Things were no better at Hethel, so there must have been a collective sigh of relief when Chrysler UK contacted Lotus in 1978 about signing an engine deal for a new rally car. At the time, Ford's Escort RS, with its 240-250bhp two-litre twin-cam, was the all-conquering rally car and Des O'Dell, Chrysler's competition director, was determined to knock the RS off its pedestal. Although Chrysler had an ideal car in the three-door Sunbeam hatchback, it lacked an engine and indeed, the time, money and facilities to develop one. It quickly dawned on O'Dell that the Lotus unit would fit the bill nicely.

A rally-tuned Lotus engine and five-speed ZF gearbox were engineered into a prototype Sunbeam which, after some tuning and testing, went on to gain second place in that year's Mille Piste rally with Tony Pond at the wheel. Encouraged by this result Chrysler gave the go-ahead for 400 Sunbeam-Lotuses to be built for homologation. However, the dealers had different ideas and such was their collective enthusiasm that plans were laid to build 4500.

Prior to the car's launch in March 1979, Chrysler had sold its ailing UK operation to the French Peugeot-Citroen group so the new rally weapon was rechristened Talbot Sunbeam-Lotus.

Happy as he was with the engine's performance, O'Dell demanded more torque and persuaded Lotus to take the engine capacity out to 2.2 litres by increasing the stroke to 76.2mm. Concerned that this would set up bad vibrations, Tony Rudd developed a flexible flywheel which dampened out the problem and allowed the engine – re-designated Type 911 – to produce 150bhp at 5,750rpm and 150 lbs/ft torque at 4,500rpm. In rallying trim the 2.2 produced 250bhp, sufficient for Henri Toivonen to win the 1980 RAC Rally with team mates Guy Frequelin and Russel Brookes third and fourth, respectively. But this was just a prelude of things to come: in the following year works Sunbeam-Lotus won the Argentine Rally and came second on the Monte Carlo, Portugal, Corsica, Brazil and San Remo events, sufficient for them to win the prestigious Championship of Makes and for Frequelin to finish as runner-up in the Drivers, series.

Despite all this success only 2,298 Sunbeam-Lotus were sold between 1979 and 1981, half the number originally envisaged. The Type 912 2.2-litre Lotus used in its own cars from 1980 onwards differs from the rallying version in having, amongst other things, a redesigned sump and main bearing panel; it also produced 160bhp at 6,500rpm and 160 lbs/ft torque at 5,000rpm.

If there is one engine that means 'Lotus' as much as an air-cooled flat six is a 'Porsche', it's the Type 910 turbo, nowadays the only engine Hethel builds. The Esprit with its aggressive Giugiaro styling always deserved more than 160bhp and while many at Lotus fought for the Type 908 four-litre V8, this project was abandoned after a Lotus Engineering client pulled out of a deal to buy sufficient engines to make manufacturing economically viable. Turbo-charging, which was rapidly becoming fashionable in 1980, seemed the way to go, but only if Lotus could conquer the dreaded turbo-lag problem which cursed blown engines at that time.

The Type 910 isn't just a normally-aspirated engine with a turbo bolted on the side, it is a fully re-engineered power unit that has been progressively developed into one of the world's great turbo-charged engines.

Although based on the 2.2-litre 907 there are a number of significant changes: lowering the compression ratio to 7.5:1 from 9.5:1 was a major step towards improved throttle response, but also entailed designing new forged pistons while new camshaft profiles opened the valves further and longer. Inlet valves were sodium filled, the valve seats hardened, cylinder head water passages enlarged and a bigger water pump installed to help reduce temperatures. The integrated main bearing panel was also strengthened and to ensure there was adequate oil supply a dry sump system with a a separate oil tank, an additional scavenge pump and an oil-cooler installed. (This system lasted until March 1983, when the Turbo Esprit inherited the radial sump baffle system originally designed for the stillborn V8).

A single Garret AiResearch T3 turbo supplied eight pounds of boost

Above
The source of all that power, the charge-cooled Turbo engine developed in-house at Hethel. (CTP)

Right
The Esprit Turbo engine is a tight fit and access for servicing is strictly limited, this is an American specification fuel-injected model. (Focalpoint)

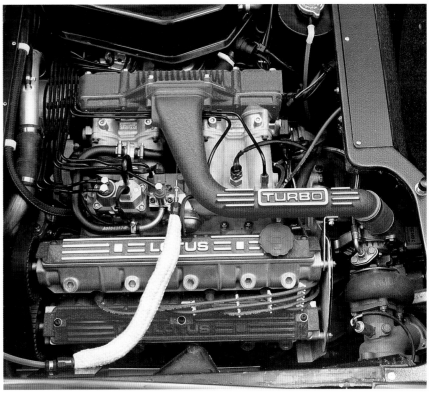

to the pressurised Dellorto 40 DHLA twin-venturi side draught carburettors. Never ones to conform, the Lotus turbo system blew rather than sucked air through the carburettors which meant special seals for the throttle spindles to prevent the petrol/air mixture escaping.

Unconventional the Turbo Esprit's powerplant might have been, but you couldn't argue about its effectiveness: 210bhp at 6,500rpm and 200 lbs/ft torque at 4,250rpm, enough to slingshot *Motor's* road testers from 0- 60mph in 5.6 secs and on to a 140+mph maximum, though the latter was somewhat down on Lotus' optimistically-claimed 150+ top speed. The turbo engine remained unchanged until early 1987 when the HC version debuted with its compression raised to 8.0:1 and boost increased to 9.5 psi. Those changes – together with bigger Dellorto 45M DHLA carburettors, a smaller turbine for better low-speed response, a balance pipe for more even mixture distribution, enlarged exhaust and inlet valve passages and more efficient oil and water cooling systems – added a further 5bhp at 6,000rpm and, more significantly, an additional 20 lbs of torque at the same 4,250rpm engine speed. On the test track this lopped a fifth of a second off *Motor's* 0-60mph time, though the HC's maximum speed was still 6mph shy of the magic 150.

For 1987 Lotus concentrated on developing the car's looks and handling rather than its performance, nevertheless the blown four got a watercooled turbocharger and an integral, rather than separate, wastegate although neither improvements affected the bhp and torque delivered. The normally-aspirated engine – Type 912S – also got a power hike to 172bhp and 163 lbs/ft thanks to a 10.9:1 compression ratio and a pair of Dellorto DHLA 45 D carbs.

At this time the search for ultimate power didn't force the pace of Lotus' engine development, but rather a determination to maintain power outputs in the face of increasingly stringent emission and fuel consumption regulations from the USA, which was becoming a vital market for Lotus sales.

GM's takeover of Lotus in 1986 gave the engineers at Hethel access to possibly the world's largest parts bin – suddenly there was scope to purchase and use sophisticated engine management systems and the like at attractive prices. In the past Lotus' small production numbers meant that original equipment suppliers were reluctant to provide Lotus with the components at realistic costs.

To meet the US 'Federal' regulations the Esprit Turbo had to go fuel-injected and, thanks to a multi-point system from GM subsidiary, AC Delco, Lotus were able to meet the exhaust and fuel consumption regulations at a lower cost than the previously used Bosch K-Jetronic system.

Ironically, the US-specification cars produced more power – 228bhp

How much more power can Lotus extract from its venerable 2.2-litre engine? In Sport 300 guise a re-worked cylinder-head, hybrid turbocharger, an enlarged chargecooler and revised engine management system pushed output up to 300bhp giving the car a 167mph top speed. (Focalpoint)

at 6,000rpm – and torque, 218 lbs/ft at 4,000rpm, than either European or UK cars. Although both markets eventually inherited this system when the non-turbo, carburetted, Esprit was discontinued from the model lineup. Using an advanced MPFI and engine management system opened up tremendous possibilities for Hethel's engine boffins – the pursuit of raw power for power's sake was now on.

In May, 1989 the Esprit Turbo SE debuted with some startling figures: 264bhp at 6,500rpm, 261 lbs/ft at 3,900rpm, 0-60mph in 4.7 secs, 0-100mph in 11.9 secs and a 164mph maximum speed. The Esprit had become a true supercar.

These figures weren't generated by a new engine, but a highly refined and developed version of the venerable 2.2-litre straight-four, making it the most powerful 16-valve engine in the world at 121.5bhp per litre. The only major internal change to the engine were Mahle-forged pistons with chrome-plated crowns running at an 8.0:1 compression ratio, although items like the induction system and the (catalysed) exhaust were updated. The massive power increase was mainly due to a sophisticated MPFI system and a charge-cooler.

The MPFI system was controlled by a fully adaptive Delco Electronics ECM which 'learnt' the precise engine demands for each journey, according to ambient temperatures, engine loads, fuel used etc. as well as controlling the quartet of Multec fuel injectors. The injectors normally pulsed once every engine revolution, with half the fuel requirement for each cylinder's combustion being supplied by each pulse ('alternating double fire'). At very low injection rates the system automatically changed to alternating single fire to maintain a minimum injection period for accurate fuel flow. At very high engine loads additional fuel flow was provided by two secondary injectors positioned in the plenum nozzle.

A liquid-cooled charge air-cooler was selected to reduce the turbo-charged air temperature and, to put it simply, ram more into the combustion chambers. The engine-mounted charge-cooler connected the turbo compressor directly with the plenum nozzle. Consisting of a series of liquid and air passages with internal finning, the cooling liquid was circulated through the cooler by a pump driven off the distributor shaft and thence to a front-mounted radiator. This system was independent of the engine's own coolant network.

The conventional coil and distributor were replaced by a direct ignition system with two separate ignition coils, while a block-mounted knock sensor retarded the spark if detonation occurred. If the detonation continued for more than two seconds then boost pressure was automatically reduced.

Under normal circumstances the Garrett TBO3 turbo blew at 12.4psi, but during hard acceleration below 3,000rpm overboost cut in for up to 30 seconds for even more rapid progression. The new Esprit duly stunned the motoring press. Richard Bremner wrote in Car; "In the roll-call of supercars, the Lotus Esprit Turbo can suddenly stand proud. It now has the horsepower and performance not merely to stand comparison with Porsche 911s and Ferrari 328s, but to embarrass them, too."

For MY '91 Lotus went all-turbo, dropping the normally-aspirated base Esprit from the UK lineup. The mid-range turbo was, in fact, the fuel-injected 2,28bhp model that the Americans and Europeans had enjoyed for some time. Following the introduction of the charge-cooled Esprit, Lotus undertook a racing programme in the American SCCA World Challenge Series for sports cars. In addition to the lighter bodyshell, the competition SE's engine power was further increased to 300bhp thanks to a reworked cylinder head, a larger Garrett T4 turbo,

The still controversial DeLorean DMC12; the rear-engined, gull-winged coupe which Lotus spent two years developing for John DeLorean between 1978 and 1980. Production lasted from December 1980 through to October 1982 when the receivers were called in. More than a decade later, the financial shenanigans involving DeLorean, Colin Chapman and Fred Bushell (Chapman's right-hand man ever since the days at Hornsey), and others has never been satisfactorily resolved

larger injectors, a recalibrated engine management system and a bigger charge-cooling system created by integrating it with the redundant air-conditioning radiator. The power increase was impressive: 300bhp at 600rpm, an estimated 261 lbs/ft at 3,900rpm, a claimed 168mph max and 0-60mph in 4.4secs.

Eventually this engine found its way into the limited edition Sport 300 which appeared at the 1992 Birmingham Motor Show. Despite the engine's advanced years and growing press criticism – "the Esprit's snarling slant-four sounds hard-edged and coarse, even metallic when idling" (Roger Bell in *Car*) – Lotus reaffirmed its faith in the 22 year-old engine when, in early 1993, they revealed the Esprit S4, simultaneously announcing that the 264bhp engine benefited from all new castings for the block, head and sump for increased stiffness and cooling capacity. The Turbo's engine management system was also re-mapped for improved response and driveability, although power and performance figures remained unchanged over the original charge-cooled Esprit. A two-litre, tax-breaking version producing 186bhp was also built for the Italian, Greek and Portuguese markets.

As the years slip by and more rival sports car producers abandon the turbo concept in preference to six- or eight-cylinder normally-aspirated engines, it is difficult to see just how Lotus can maintain its present position with a small-capacity four which, for all its sophistication, has to work very hard for a living. Perhaps Lotus will have to turn to a major supplier for its next engine block and develop their own heads etc. If that transpires then the circle will have been completed, but *en route* will be the legacy of a truly fine engine that became a benchmark for prodigious specific output from a small capacity.

Elite Series 1 to Eclat Series 2

If the 900 engine series was going to be the heart of Lotus production models to come, then Chapman had equally exciting plans for the cars themselves anticipating a comprehensive model range:

M50 Four-cylinder, four-seater
M51 V8-powered, four-seater
M52 Four-cylinder, 2+2
M53 V8-powered, 2+2
M70 Four-cylinder, mid-engined two-seater
M72 V8-powered, mid-engined two-seater

Chapman was certainly ambitious for Lotus, with visions of producing cars ranging across the price and power spectrum to rival great European marques like Ferrari and Mercedes-Benz. It was heady stuff that Chapman dreamed of and with increasing F1 success it must have seemed that he possessed the Midas touch – from kit-car maker to Mercedes and Ferrari rival in less than four decades would appear to have been the ultimate post-war motoring legend.

When Tony Rudd joined Lotus in 1969, John Frayling had already penned an M50 design which was undergoing wind tunnel tests at MIRA. Neat, though the styling was Chapman felt it too conventional for the 1970s and instructed designer, Oliver Winterbottom to come up with another concept.

The original plan was for the new car to appear in 1972 and supersede the discontinued Elan and Lotus 7 which had fallen by the wayside. However, a decline in Group Lotus profitability in '70 and '71 restricted expenditure on the new car which meant it didn't appear until 1974 – even though its design had been approved in April 1971.

If approving a design in April and hoping to have it in production the next year seems a fantasy, that is to ignore the pace at which Lotus worked 20 years ago. Despite it being an evenings and weekends job for the principal players involved – Rudd, Kimberley and Winterbottom – a styling model had been completed by September '71; a running

The Lotus Elite was the marque's final break with its past as a kit-car producer although it still retained basic Lotus principles such as a back-bone chassis and glass-fibre bodywork. The Elite was designed to be more than a just a sports car; it was seen by Lotus as a four-seater Grand Tourer, though the rear passengers would need to have been vertically challenged to enjoy trans-continental trips. (Monitor)

prototype was commissioned four months later plus a further two in April 1972, so it is feasible that production could have commenced in the following autumn, 18 months after the project started.

That it didn't was due to forces outside of Hethel and beyond even Chapman's control and influence. Legislators in both Europe – the UK was now a member of the EEC – and the US was demanding ever more stringent emission and fuel consumption legislation and, especially, accident protection. As a consequence, the M50's design was being constantly updated to comply with the new rules. As Chapman became increasingly frustrated with the legislation which he felt was comprising the concept, his eye was taken off the ball by a young Italian designer.

International motor shows are the industry's fertile hot houses; neutral meeting grounds where manufacturers, engineers and designers plant the seeds of new ideas. Most wither and die – but provide good copy for the motoring press – only occasionally bearing fruit, but when they do it can be spectacular.

Imagine then, the situation at the 1971 Geneva Salon when Italy's brightest up-and-coming stylist, Giorgio Giugiaro suggested to Colin

Chapman that he create a concept car based on a Lotus chassis.

Giugiaro had just established ItalDesign and had yet to gain status as one of the world's most influential post-war car designers. Even so, he had achieved universal praise and recognition for stunning cars such as the Alfa Romeo 2600 Sprint, Iso Grifo and Fiat Dino Coupe to name but three. What could be more natural than for one of the world's most exciting stylists to combine with one of the greatest automotive engineers to produce a dream car? Chapman was flattered and since Lotus' own M70 project was on permanent hold he saw it as a way of realising his vision of a Lotus family of cars that much sooner. M70 was conceived with a 98ins wheelbase, seven inches greater than the Europa's, as well as wider front and rear tracks – up six inches from 53.5ins – to provide more interior space than the cramped Europa.

Therefore in mid-1971, a heavily modified Europa Twin-Cam chassis was shipped to ItalDesign's styling house on the outskirts of Turin for work to begin. Never one to stand still it was about this time that Chapman became interested in buying Aston Martin. Concerned that the M50's purchase price (and therefore the type of buyer it would attract) was beyond Lotus, Chapman reasoned that a traditional luxury marque such as Aston Martin would have more success at selling the car. However, after an initial foray to Aston's Newport Pagnell factory, Chapman decided against the purchase and to continue Lotus' move upmarket independently.

Even while the M50 programme was taking shape, Giugiaro's team was working on the dream car. Early renderings betray a strong resemblance to the Maserati Merak and Bora – which Giugiaro also penned – but gradually their softer curves disappeared, being replaced by the hard crease lines which became Giugiaro's trademark for the next decade. At the November 1972 Turin Motor Show a full-size mock-up was ready for display. Standing alongside the Maserati Boomerang concept car, the Lotus was an instant hit with the media and public alike.

So impressed was Chapman that he gave the go-ahead for further development, with Kimberley and Winterbottom being given the task of turning the 'Silver Car' prototype into reality. For the next year this duo virtually lived in Turin working with ItalDesign and Chapman himself flying out to Italy two or three times a week to supervise the project.

By the end of 1973, the first running prototype was back in the UK where Colin Spooner, later to become Lotus' design director, continued working on the project at Ketteringham Hall, a country house a mile or two from the Hethel factory and now home to Team Lotus. Meanwhile, the Elite's production and launch was fast approaching, scheduled for March 1974. However, the debut was beset with problems: production costs had risen and Chapman's vision of selling 50 £3,500 Elites a week

Right
Chapman's master plan for a range of sports cars to rival Europe's best included a mid-engined two-seater to replace the Europa. The answer was the dramatically styled Esprit, penned by the then up-and-coming Giorgio Giugiaro. Once the Esprit got into series production there was an easily identifiable Lotus look: all three cars were based round a backbone chassis, were made from glass fibre and had a distinctively wedgy style to them as well as sharing a common powerplant and other componentry. (Focalpoint)

Below right
This side elevation displays Giugiaro's origami styling and the Esprit's aggressive arrow-like profile. (Monitor)

Above

A year after the Elite, Lotus produced the Eclat. Intended as a 'cheap' version of the Elite, its styling was more successful with the vertical rear glass hatch replaced by a sloping roof line that led into a conventional boot. In years to come it would be this derivative that would stand the test of time. (Focalpoint)

Left

This Eclat's interior with its black suede fascia, chrome trim and the red leather upholstery reflects the fashions of the time — note how little knee room there is between the passenger's seat and the rear bucket seat. (Focalpoint)

had been savagely pruned to 10 cars a week at £5,000 each.

To make matters worse, the miners went on strike in February 1974 and the Government imposed a three-day week. Unfortunately, Lotus had already booked its media advertising campaign and in March the *Sunday Times* appeared with an advertisement for the new Lotus, nearly two months ahead of the car's revised launch date.

Although the new Elite retained the traditional Lotus backbone chassis and GRP bodywork it was a total departure from its predecessor's kit-car background. Power came from a pukka Lotus version of its new slant four engine, producing an additional 15bhp to that supplied for the Jensen Healey. Despite this, test cars failed to achieve Lotus' claimed 128mph max – *Autocar* could only manage 124mph – though the same magazine did equal Lotus' 7.8secs for the 0-60mph dash.

In Lotus terms the Elite wasn't a very light car at 2,598.4lbs, a hefty 672lbs heavier than the Elan +2S and, remarkably, 56lbs heavier than the all-steel Austin 1800.

Mechanically the car didn't really break any new ground: the engine was already running and it was now mated to an Austin Maxi-derived five-speed gearbox. The suspension was all-independent with coil springs, wishbones and an anti-roll bar at the front while the rear bore allegiance

to the Europa with fixed-length drive shafts, lower transverse link and a long pressed-steel semi-trailing radius arm hinged from the chassis.

Braking, servo-assisted, was by 10.4ins discs at the front and, unusually, inboard-mounted 9x2.25ins drums at the back. The smart seven-inch alloy wheels were die-cast by GKN and fitted with Dunlop Supersport 205/60 VR 14 tyres.

The most startling aspect of the car was Oliver Winterbottom's wedge-like design. With its ground-hugging 0.3Cd styling and tailgate access to the small boot it combined elements of contemporary F1 aerodynamics with on-road practicality and the ability to carry (just) a quartet of adults in its Giugiaro-penned interior.

However, the body style also contained a hint of something far more radical than slippery aerodynamics. For years Chapman had been developing a revolutionary new glass fibre moulding process with which

Above

You could have the UK version of the Eclat Sprint in any colour you liked as long as it was white with a black stripe on the bonnet and sides. Power came from a standard 160bhp version of the Lotus twin-cam, but five-speed models had lower final drive ratio while four-speed variants had new 5.5ins wide wheels. (Focalpoint)

Left

This Eclat Sprint for the States – where it was known as the Lotus Sprint – is not to be confused with the later 'Sprint' model which was sold in the UK. (Focalpoint)

The late, great Ronnie Petersen with his Esprit S2; note the neater treatment of the front spoiler. (Monitor)

to build his luxury cabin cruisers (he had bought both Moonraker boats and JCL Marine with his own money). Until now, both glass fibre cars and boats had been made using a labour-intensive hand-laying process. Chapman developed the Vacuum Assisted Resin Injection system in which pre-cut glass fibre matting is laid into a mould and, as the resin mix is injected, the air is drawn out.

The Elite was built in two halves – top and bottom – using this system, hence the car's pronounced midriff rubbing strip which hides the seam. Although two models, the 501 and 502, were announced at the time of launch – the 501 was the basic model and the 502 came with air-conditioning, quartz-halogen headlamps etc – the base model didn't go on sale until January 1975 by which time the 503 – fitted with power steering, electric windows etc as standard – was already available.

The principal motoring magazines were given a brief taster of the Elite's capabilities on the Hethel test track during the May launch, but it wasn't until January 1975 that *Autocar* got a 503 for full appraisal, concluding that the car had "typical Lotus excellence of handling combined with good ride. Good performance from Lotus...engine if

Other key differences with the Esprit S2 were black plastic ducts behind the rear quarter windows and a much tidier arrangement under the rear bumper. (Monitor)

revved; rough at top end. Not quiet enough, but a highly enjoyable and relaxing mile-eater."

The final Elite, the 504, appeared in October 1975 fitted with Borg-Warner's Model 65 three-speed automatic transmission. Costing only £30 short of £8000, it was the most costly Lotus to date and never really attracted the hoped-for sales.

Unfortunately for Lotus, it couldn't have launched its new range of high speed GTs at a worse time. Not only was the UK suffering from the after effects of the early 1970s industrial anarchy, but the world was reeling from the Yom Kippur war and massive increases in oil prices. High-performance sports cars were not in vogue, yet that is all Lotus built and, to Chapman's credit, he didn't flinch but pressed on with developing the Giugiaro-inspired dream car which, by now, had been christened 'Esprit'.

Colin Spooner's team had been working flat-out to ready the Esprit for its world debut at the 1975 Paris Salon, a few weeks prior to the Earls Court Motor Show where it would be seen by the British public for the first time alongside yet another new Lotus. The show car's box

section chassis might have been cobbled together from a Europa, but the genuine article differed by having a broad cradle at the back to house the slant-four engine and, more importantly, the V8 which Chapman dreamed would one day power the Esprit.

Much to Mike Kimberley's consternation, Chapman chopped two inches out of the Esprit's 98-inch wheelbase. This might have made the car visually more appealing and compact, but it meant tall drivers found the Esprit's cabin too cramped – a legacy the car would always live with. Not only that, but there was virtually no storage space in the small cabin and only soft bags could be carried in the stowage area at the rear of the mid-mounted engine.

The other major styling problem was the windscreen's acute rake, a meagre 22 degrees from the horizontal and well beyond the permitted minimum. Chapman was determined to retain the car's arrow-like profile and, after hours of burning the midnight oil with Giugiaro and Kimberley in the Turin studios, finally achieved a solution; the 'A' pillars remained as they were but, the windscreen's centre line was pushed back until it was practically flat and sloping at the same 26 degree angle as the 'A' pillars.

In 1980, with the ensuing DeLorean crisis still to come, Chapman had other things on his mind: the first true Lotus supercar, a turbocharged version of the Esprit, capable of nearly 150mph and 0-60mph in a shade more than six seconds. This is the original Turbo Esprit before it gained a more familiar livery

A Lotus test driver giving an Essex Turbo Esprit some serious stick round the Hethel test track. (Focalpoint)

The major mechanical difficulty was trying to find a suitable replacement for the Europa's Renault-derived transmission. Eventually rival French manufacturer Citroën provided the solution by offering Lotus its own five-speed transaxle.

Developed from that used in the DS series, this unit was employed in Citroën's futuristic front-wheel drive SM and, later in the mid-engined Maserati Merak. A conventional two-shaft design with output to the spiral bevel gear from the second shaft, its layout meant the crownwheel and pinion and final drive casing could be used in both front and rear-wheel drive cars. There was also a choice of gear and final drive ratios, though Lotus opted for the same ones used in the SM coupe and the original Meraks.

Suspension followed typical Lotus thinking with the transverse and semi-trailing lower links acting as a wide-based wishbone and the drive shaft doubling up as an upper suspension arm. Double wishbone front suspension incorporated Opel Ascona components with combined coil spring/damper units all round.

Solid Girling disc brakes were fitted, with those at the back mounted

inboard while steering was rack-and-pinion. Wolfrace cast alloy 14-inch wheels were fitted with six inch rims at the front, whilst those at the back were an inch wider.

Unlike the Elite, the Esprit was hand-laid as the bodywork couldn't be designed in two separate pieces for the VARI moulds. However, it was self-coloured which helped to reduce labour costs, but restricted the number of available colours.

Powered by the standard Lotus two-litre engine, the S1 was launched with a £5,844 price tag. However, this proved premature as development time dragged on until May 1976 by which time it had increased drastically to £7,833. This might have been acceptable if the motoring press had raved about the car's qualities. Unfortunately they didn't: roadholding and handling were universally praised, but the press were equally damning about lack of steering feel, visibility (always an Esprit failing), build quality and especially performance. Lotus claimed a 138mph maximum, but none of the press cars could be coaxed past 130mph.

The excitement created by the Esprit's UK debut eclipsed another newcomer which was to play a far more significant role in Lotus' fortunes than Chapman ever intended. On the face of it the Eclat was nothing less than a bargain basement version of the Elite. Chapman's model plan had called for a four- cylinder 2+2 – M52 – and the Eclat was

Above
The Lotus VARI system was developed by Chapman to speed up production of glass fibre products as well as to improve quality. Essentially the various mattings and foam beams which constitute the body shell are hand-laid in the moulds, then resin is injected in as the air is drawn out. The body is then left to cure, before being extracted and taken away for assembly

Above right
Once the two halves of the body shell have been joined, they undergo intensive fettling to ensure the best possible finish

Right
Final assembly; building a Lotus always has been, and still is today, a very labour-intensive process

supposed to be just that but, in fact, the car's elegant fastback styling, again the work of Oliver Winterbottom, meant that to all intents and purposes it was still a four-seater.

The only external difference between the Eclat and Elite was the abbreviated roofline which sloped downwards from the 'C' post to the tail lights rather than extending horizontally to finish in a hatchback. Winterbottom's clever styling meant virtually no headroom was lost, boot space was improved and, most importantly to a cost-conscious operation like Lotus, only minimal changes had to be made to the upper body mouldings.

In an effort to take more costs out of the vehicle and therefore make it appeal to a wider band of prospective buyers, the basic 520 version used the cheaper Ford Granada/Capri four-speed gearbox rather than the hybrid Lotus/Austin Maxi five-speeder while the Elite's 4.1:1 final-drive ratio was replaced by a 3.73:1 differential. Out, also, went the original GKN alloys replaced by skinny, 5.5ins, steel rims shod with measly 185/70HR 13ins rubber.

At its launch the Eclat 520's 5,729 asking price undercut the nearest Elite's by a full £754. Over the coming months Lotus launched a full Eclat range which mirrored the Elite's specification levels with the 502 matching the 522 etc.

Although aimed at those with less spending power, the lower spec'd Eclat never sold. Customers preferred the better-equipped models. Not only did the Eclat attract buyers but favourable press comments too. *Motor* eventually got its editorial hands on a 523 and declared: "Outstanding performance and economy, excellent five- speed gearbox. Remarkably fine road-holding…quiet and refined. Completely fulfills the Lotus prestige-car aspirations."

Fulsome praise indeed, with 0-60mph now accomplished in 8.5secs and a maximum speed of 124.3mph recorded. No doubt some of these performance improvements were attributed to the revised 160bhp engine, but a lot must be down to a 120lbs weight reduction thanks to the re-style.

A 'Sprint' version of the Eclat appeared in 1977, though its badging was something of a misnomer as, apart from lower final-drive ratio, 4.1:1, on the five-speed models it was mechanically identical to other Eclat and Elites. Only available in white, the Sprints could be distinguished by black side stripes, a matching bonnet flash, bootlid and filler cap. There was also an unmissable Union Jack/Sprint decal on the car's rump.

No, it's not your eyes but a picture of an active-suspension Esprit doing its party tricks outside the Lotus factory. Lotus first started work on this technology in mid-1981 after its controversial twin-chassis F1 car, the 88, was banned. One of their early test drivers was Nigel Mansell who eventually went on to win the 1992 F1 World Championship in a Williams fitted with a type of active suspension. (Focalpoint)

Stung by media criticism of the Esprit, Lotus swung into a hasty development programme which resulted in the S2's launch in August 1978. There were no blinding flashes of inspiration with the S2, it was more a case of making the car what it should have been in the first place. Visually there was little new apart from a proper wrap-around front spoiler, air scoops behind the rear windows, Rover SD1-derived tail lamps and Speedline road wheels.

Although there was nothing Lotus could do to enlarge the interior it was tidied up with a new instrument cluster, switches, re-shaped seats and a digital clock. Thanks to the 160bhp engine, *Autocar* managed to hit 130mph and zero to 60mph in eight seconds when it tested the car. Ride and handling were better thanks to subtle changes in front wheel offset, damper settings and the steering column bushes.

A hundred limited edition Esprits were built in celebration of Mario Andretti's Formula One World Championship in 1978. Garishly decked out in black and gold JPS colours, each car bore a facia-mounted plaque signed by Chapman himself. As the 1970s drew to a close, Chapman spent more and more time with his beloved racing team, distancing

himself from car production, so that by 1977 Mike Kimberley had been elevated to managing director of Lotus Cars. Around about the same time a five year loan worth £2.2 million was negotiated with American Express as a means of recapitalising the £6 million Lotus had borrowed to develop the Elite.

This struggle for survival had meant that the Elite, Eclat and Esprit hadn't undergone any real updates or developments since their respective launches. However, thanks to Chrysler's order for the 2.2-litre Sunbeam-Talbot rally special, Lotus were able to offer a re-vamped series of cars for 1980. Lotus launched itself into the new decade in the biggest

Above
During this period, Lotus produced a bewildering array of special editions. This is a Riviera Éclat from 1980-2 with a lift-out roof panel and a vestigial rear spoiler. (Focalpoint)

Right
As the years progressed the interiors became more sophisticated and tasteful. This is a post-Chapman era car as can be seen from the Lotus lozenge on the seat rake adjuster. The black bulb next to it is to inflate or deflate the seat's lumbar support. (Focalpoint)

possible way with a party at the Royal Albert Hall, a million dollar bash paid for by Team sponsor Essex Petroleum.

Back in 1978, Chapman had met Essex' chairman, David Thieme at a sponsor's lunch during the weekend of that year's Monaco Grand Prix. A hugely successful oil-broker – his 1979 profits were estimated at US$70 million – Thieme was immediately attracted to the idea of getting involved with Formula One. Coincidentally, John Player decided to quit motorsport sponsorship and the following year the Lotus F1 team appeared with backing from the Martini drinks firm and Essex logos. For 1980 Martini disappeared and Essex took over as prime sponsor. David Thieme had great plans for his company which included the Indianapolis 500, Le Mans 24-hour race and a road car or, at least, an 'Essex-Lotus'.

A thousand guests – fed by the 'Moulin de Mougins' star chef, Roger Verge and entertained by Shirley Bassey – witnessed the most extravagant car launch there's possibly ever been, especially for a company so small as Lotus.

From clouds of billowing dry ice emerged two single-seaters – a Lotus 81 F1 car and a Indy 500 Penske – flanking the new Essex Turbo Esprit resplendent in its sponsors' garish blue silver and red livery. With a claimed top speed of 152mph and 0-60mph in under six seconds, the Turbo Esprit (it didn't become an Esprit Turbo until 1987), was Lotus' first supercar and the start of the Esprit legend. Visually, the car bore allegiance to the original Esprit, but Giugiaro's addition of a deep front air dam, flared sills with NACA ducts and extended rear valance gave the previously delicate-looking Esprit real road presence. The sloping rear window was abandoned in favour of a Venetian blind-like cover with a small spoiler at the topmost edge designed to create a low-pressure area which would suck hot air from the engine compartment. A deeper spoiler was added to the trailing edge of the boot. (A vertical rear window now separated passenger and driver from the engine compartment).

The wheelarches were crowded full of fat Goodyear NCT tyres – 195/60 VR15 at the front and even bigger 235/60 VR15s at the back – mounted on BBS spoke-pattern wheels, although the original Essex cars came on three-piece Compomotive alloys. The Essex Turbo Esprit might not have looked particularly subtle, but its image of having just come off the race track wasn't a false promise.

To cope with the turbocharged engine's prodigious power and torque – 210bhp and 200 lbs/ft – the chassis was totally redesigned. Torsional rigidity was claimed to be improved 50 per cent thanks to increased use of box and triangulated sections. The original Opel-based front suspension was discarded in favour of upper wishbone and lower transverse link, derived from the Elite/Eclat, and the practise of using the

driveshaft as the rear suspension's top link was abandoned at last. A separate transverse link took its place and the driveshaft was fitted with a plunging velocity joint to cope with the new geometry, and the engine was located on four-point mountings to reduce NVH. Intriguingly, the engine's tubular cradle was much wider than needed for a turbocharged straight-four and, although Lotus denied it for years to come, it quickly became common knowledge that there were plans for a V8-engined Esprit (M72 against the Turbo's M70 moniker) and, although a 330bhp prototype was built and run, the car never went into production.

Braking performance was uprated with 10.5ins and 10.87ins solid discs front and rear respectively. Those at the back were mounted inboard adjacent to the unchanged Citroën transaxle.

The interior remained basically the same, but was trimmed out in rouched leather and included air-conditioning and a complicated Panasonic stereo system mounted in the central roof panel. These 'extras' were included in the first 100 Turbo Esprits built — all in Essex colours — and reflected the car's £20,950 asking price.

Although development of this car had been underway since 1977 — but delayed by the DeLorean project — sales didn't start until August, 1980 by which time a new range of Eclats and Elites had debuted.

Tagging the models S2.2 reflected use of the larger 2200cc 912 engine which, although it didn't boast any more horsepower than the 907, produced an extra 20 lbs/ft of torque and, more importantly in terms of driveability, created 140 lbs ft (the 907 engine's peak output), at only 2400rpm.

The other major mechanical changes included a hot-dipped zinc coated chassis to reduce rust and adopting a Getrag 5-speed gearbox in place of the more costly Lotus-built transmission with a 4.1:1 final drive ratio as standard and a higher 3.73:1 offered as an option. Lotus also persisted in offering Borg-Warner's three-speed auto for an extra £260 but customer take-up was negligible. Minor styling changes included an Esprit S2-like front spoiler, under sill extension and Rover SD1 rear lights incorporated into a new bumper design. Apart from some new trim details, the S2.2's interior was virtually unchanged from its predecessor.

The normally-aspirated Esprit also received the enlarged 2.2 litre engine and galvanised chassis, but apart from a new front spoiler, black trim on the rear windows and re-styled wheels, the S2.2 was little different to the old Esprit. However, the S2.2 was to prove a short-lived

Don't be confused by the Eclat badging, this is an Excel; for a period the car was known as the Eclat 3, and could even be classified as an Eclat Excel. Look carefully and you can see subtle styling changes round the bumpers and the rear window, while Toyota door handles replace the Morris Marina ones previously used. (Focalpoint)

model – only 46 built – before it was replaced in April, 1981 by the much-improved – and cheaper (by £1,809 at £13,461) – S3. At the same time, Lotus also started offering non-Essex Turbo Esprits in standard body colours, with leather and air-conditioning as optional extras, at the more realistic price of £16,917.

The S3 was a much-improved car: it gained the Turbo's stronger and galvanised chassis, revised suspension and bigger brakes while the larger – 15ins – wheels could be specified as an option. Even the bodyshell inherited some Turbo styling with new bumpers, re-vamped engine bay and air intakes immediately aft of the rear quarter windows. To distinguish it from the more powerful Esprit, the S3 did without the additional spoilers and sill extensions and retained the large rear tailgate/window assembly.

The Esprit's appearances in the two James Bond films – *The Spy Who Loved Me* in 1976 and *For Your Eyes Only* in 1981 – had, perhaps, hyped the cars beyond customer expectations, but with the Turbo and S3 now available, Lotus were able to live up to those promises. *Autocar* eventually got hold of a Turbo and an S3 to test within a month of each other – May and June '81, respectively – and whilst these independent acceleration figures might be shy of Hethel's claims, both cars proved to be credible performers with the S3 recording a 134mph max, 14 less than the Turbo, and 6.7secs to 60mph compared to the Turbo's 6.1 secs. Rival magazine *Motor* broke the six-second barrier with its Turbo recording 5.6secs, but its 140mph max was well shy of Lotus' claimed 150+mph top speed.

In spite of all these improvements to the model range, Lotus sales were at rock-bottom. The UK was in the grip of an ever-worsening recession and the last things on people's shopping lists were costly GTs. Even limited edition specials like the Riviera Elites and Eclat, complete with sunroof, and the Elite winning the prestigious Don Safety Trophy failed to revitalise sales. It wasn't until Lotus took the desperate step of slashing their prices – by reducing their own and dealer margins – that sales started to recover: in October, Eclat prices came down £1,894 to £14,857 and from January 1982 the Elite was £1,616 cheaper at £15,590.

Much of these problems can be attributed to Chapman's laudable but costly policy of producing as many of the cars as possible at Hethel. At one stage upwards of 70 per cent of each Lotus was built in Norfolk and, though buying in Getrag gearboxes etc might have helped, there was still

As the years passed, the Eclat gained more popularity and eventually overtook its predecessor. Before long Elite production was barely scraping into double figures and in 1982 it was dropped altogether. Nevertheless it played an important role in establishing Lotus as a more serious player in the international sports car arena

a long way to go before production costs could be reduced to help profitability. Matters weren't helped by the fact that the Elite and Eclat's styling had remained virtually unchanged since their launch seven years earlier. But as the company wasn't earning vast profits, there was little or no spare capital for investing in product development; Lotus was caught in the classic trap in which many small manufacturers have found themselves.

Chapman appreciated that costs had to come down and that meant buying in more proprietary components from outside. He could also see the advantages of a larger company taking a stake in Group Lotus. Amongst its engineering clients Lotus could include Japan's Toyota, for whom it was helping to develop the Supra's ride and handling. So, in retrospect, it isn't surprising that the Japanese took a 20 per cent stake in the group. However, it was more than just a financial interest: Lotus would gain access to Toyota components for the next generation Elite/Eclat and a 1.6-litre twin-cam destined for a new, small Lotus – the Elan of the '80s. In return Toyota would be able to tap Lotus' considerable engineering talents including the development of an F1 turbo engine, though in the end neither the Elan or F1 project came to fruition.

Development of a revised front-engined/four-seater started in Spring 1981 and made its debut in October 1982. Badged somewhat clumsily as the Eclat Excel (Eclat was later dropped), it was a considerable improvement over its forerunner in terms of looks, engineering and price as the extensive use of Toyota componentry saw the new Lotus on sale for £13,787, some £1,109 less than the Eclat S2.2.

The most notable aspect about the revised bodystyle was the cleaned up nose with its new one-piece bumper-cum-spoiler (the line of which carried on through the sills) and a tidier 'B' post treatment. A more efficient exhaust system was designed to reduce back pressure by 40 per cent and the Getrag gearbox dropped in favour of a Toyota five-speed linked to a chassis-mounted 4.1:1 Toyota final drive. The backbone chassis remained practically unaltered, apart from being galvanised and the rear suspension modified along the lines of the Esprit S3 to incorporate transverse top links and wide-based lower wishbones; Toyota driveshafts with plunging universal joints were also fitted.

The Japanese also provided the ventilated disc brakes now fitted all round (instead of the old disc/drum set-up), and moved outboard at the back while the wheel and tyre combination remained unchanged from the Eclat. Inside, the Excel Eclat was virtually identical to the Eclat though customers could specify a Momo three-spoke steering wheel in place of the standard, and rather *passe*, two-bar one. Eventually the Momo would be fitted as a matter of course.

By discontinuing the Elite – and no hatchback version of the Excel Eclat was envisaged – Lotus was now able to concentrate on the Esprit, the re-vamped 2+2 and developing the new Elan. Orders were beginning to flow in for the new car – encouraged by some very positive press coverage, *Autocar* heralded it as "completely different and dramatically improved car" – and it looked as though Lotus' dark days were behind them.

Nothing could have been further from the truth. On 16 December, 1982, Anthony Colin Bruce Chapman died of a heart attack. He was 54.

The Toyota and GM Link

Chapman's untimely death was devastating enough to Lotus, but then came the second devastating blow as American Express withdrew its financial backing.

Lotus, now under the chairmanship of Fred Bushell, was in dire straits and there were serious doubts as to whether it would survive, even with Toyota's backing. Fortunately, by June 1983 a rescue package had been put together with British Car Auctions taking 25 per cent and merchant bankers Schroeder-Wagg 14 per cent. These were joined later by JCB – the earth-moving equipment manufacturers – with 12 per cent and 14 per cent bought by a business colleague of BCA's chairman, David Wickens.

With Wickens as executive chairman, Lotus looked secure, but he didn't endear himself to the staff – or Lotus enthusiasts – by abandoning the traditional ACBC green and yellow badge for a dark green squashed lozenge bearing only the Lotus name. It was the equivalent of Fiat replacing Ferrari's Prancing Horse with a donkey. If abandoning the traditional Lotus badge was the nadir of Wicken's time at Lotus then Etna was the apogee.

Etna was nothing less than Lotus' Ferrari basher. This was the car intended to put Hethel on the same level as Maranello. Penned by Guigiaro, the wedgy 0.9 Cd car looked as it if was doing its theoretical 182mph top speed even when parked.

The advanced carbon-fibre reinforced body hid a host of advanced Lotus Engineering-developed technology including active suspension, electronically-controlled ABS and a computer-managed Continuously Variable Transmission system which was standard; a five-speed manual with a power operated twin clutch would have been optional.

Etna's heart was the much-rumoured four-litre V8 Lotus had been developing since 1979. It was more than just a pair of slant fours joined together, more like a completely new power unit. The all-alloy 90 deg V8 weighed in at 414.5 lbs fully dressed, remarkably light given that it produced 330bhp at 6,500rpm and 294 lbs/ft at 5,500rpm. Designated Type 909 features like double overhead camshafts operating four valves per cylinder, pent roof combustion chambers with centrally-mounted

Following Colin Chapman's shockingly early death, time couldn't stand still and within weeks Lotus had been re-organised under the temporary chairmanship of Fred Bushell. By the following August the company had been re-financed with capital from, amongst others, British Car Auctions and Toyota. As BCA held the largest stake, its ebullient chairman, David Wickens assumed the same role at Hethel. It was a new beginning, full of hope

Wickens quickly and controversially stamped his authority on Lotus by ditching the famous yellow and green Lotus badge for a British Racing Green shield that looked more like a squashed lozenge. It hardly endeared Wickens to Lotus enthusiasts. (Focalpoint)

spark plugs for efficient burn characteristics and a bore and stroke of 95.3mm x 70.3mm, ensured this was a state-of-the-art engine.

Unveiled at the 1984 Birmingham Motor Show, Etna grabbed the limelight from all the other manufacturers. Giugiaro was on hand when the car was unveiled and a sales tag of £28,000 was optimistically talked about. Wickens and Kimberley were candid about the need for a partner to finance engine production so that Lotus could realise the Etna dream. Unfortunately, no deal was struck and Etna remains to this day an unfulfilled vision.

Despite the newfound funding, and a considerable increase in business for Lotus Engineering, Lotus Cars still suffered from insufficient capital to develop not only the existing Excel and Esprit, but also X100 the projected new small car. By 1985 it became obvious to the board that if Lotus was to survive its future lay not as an independent, but as the adopted member of a bigger, more financially powerful family. The question was, who?

Lotus Engineering's clients numbered several of the automotive world's big players including General Motors and Chrysler and it was to both of these that non-executive chairman, Alan Curtis broached the idea of purchasing Lotus. At the time, Chrysler was none too financially secure, but GM was thriving and Bob Eaton – who oversaw much of GM's future engineering needs – was a keen Lotus fan and, indeed, employed Lotus Engineering on advanced research and design projects.

However, whilst a possible takeover bid was being negotiated in the background, model development continued. For the 1986 model year, the Excel SE appeared complete with a veneer fascia, a Momo three-spoke steering wheel and 7x15ins Speedline alloys fitted with Goodyear 215/50 Eagle NCT rubber; a Cam Gears variable ratio steering rack which reduced turns lock-to-lock from 3.1 to 2.65 was also installed.

Externally, Lotus had done some aerodynamic fine tuning with a small bib on the front spoiler and two different bootlid-mounted spoilers – one for the SE and one for the standard cars. It was the first work that Peter Stevens had done for Lotus, though it wouldn't be the last. Inside there was a new wraparound fascia, slimmer headrests, new heating and ventilating system, a tilt-adjustable steering column, central-locking, a new wiring system and a revised headlamp lift mechanism.

When *Autocar* tested an Excel in February 1986 the magazine praised its "beautifully predictable chassis and willing twin-cam", but criticised its high £18,000 price tag and "thoughtlessly positioned ignition switch", narrow opening doors and lack of fuel injection. Still, at least the car's performance was now on par with its European and Japanese rivals, with a 131mph top speed and sub-seven seconds time for zero to 60mph.

By 22 January 1986, a deal was struck and Lotus had been 'adopted'.

Lotus Eminence

General Motors bought-out BCA, Schroeder-Wagg and JCB, the Americans effectively becoming the majority shareholder. After four months of co-owning Lotus with an arch-rival, Toyota sold out to GM – though it continued to supply components – and by October of the same year, GM had acquired a 91 per cent stake in Lotus. The deal had cost the US giant a paltry – in their terms – £22.7 million.

Lotus' future looked secure and Mike Kimberley, now chief executive officer and managing director, talked boldly of an exciting range of fresh Lotus products: a new Elan, replacements for the Excel and Esprit and a supercar – code-named M300 – bristling with leading-edge technology to replace the stillborn Etna.

At first things started to move slowly. In 1986 the only changes to the model lineup were an automatic version of the Excel – the SA – which used a Lotus version of the ZF four-speed auto and a High Compression version of the Esprit Turbo, which used the same cylinder head design as that on the Excel SE. There was a reason for this slow progress, though that wouldn't be unveiled until the 1987 London Motorfair and the Florida Motor Show opened their doors to the public for the first time.

During the Wickens era Lotus announced a number of projects, none more optimistic than Eminence. First seen at the 1984 Motor Show it was conceived as a high-speed – 160mph, 0-60mph in six seconds – six-seater, four-door head-of-state saloon. Powered by a 4-litre V8 it would feature Active Suspension and a Kevlar and carbon fibre monocoque, with armour plating an option. It never got beyond this concept sketch...

Even before GM came on the scene Peter Stevens had been working on a replacement for Giugiaro's ageing design and in November 1985 he presented Project X180 to the board. Once the GM takeover was in place a few months later the green light was given for the new Esprit to be made a priority.

Sensational though the Giugiaro design was when originally launched, it had aged badly, and its Origami-like styling was jaded and *passe*. Stevens had a difficult task in that he had to follow in the tracks of a man generally acknowledged as one the most influential designers of modern time yet retain the old Esprit's dimensions. When the revised Esprit was unveiled it was generally acknowledged – with fulsome praise from Giugiaro himself – that Stevens had done an excellent job. Aided by Julian Thomson, Stevens got rid of the flat panels and straight geometric lines, replacing them with more integrated, subtly-shaped forms.

Gone, for instance, was the totally flat windscreen, which was replaced by a mildly curved one, though that still didn't totally irradicate reflections from the top of the dash on a bright day.

The flying buttresses were still there to make pulling out of acute-angled junctions just as difficult as in the old model, but between them was a plexiglass panel on the turbo model (the non-turbo Esprit made do without that which pushed its Cd up to 0.35 compared to the Turbo's 0.34).

Despite these wholesale changes, there were still a few leftovers from the past which niggled some commentators, namely the Morris Marina door handles and the doors' narrow opening angle. Not only was the body design right up to date, but so was its construction method. Originally the Esprit had been designed for VARI construction – the same method as the Excel – but lack of investment meant that throughout its life, the Esprit was hand laid. Lotus took advantage of the redesign to rectify that. Steven's Esprit was designed from the start to be built using the VARI system; that, and extensive use of Kevlar in high stress areas made the body that much stronger to the benefit of ride and handling and vehicle durability.

While Stevens and Thomson had enjoyed quite a bit of scope in re-sculpting the car, colleague Simon Cox – who was charged with re-vitalising the interior – had a less easy task.

As the chassis was identical to the old Esprit all the gains on interior measurements came from the body and were strictly limited: 0.8ins greater headroom, one inch more footwell width, 0.6ins extra legroom and inch wider seats. Better than nothing! Out went the slightly 'tarty', rouched leather interior, replaced by a more tasteful, and durable, tweed-leather combination. The fascia, whilst paying homage to Guigiaro's original 'boomerang' design, was more cohesive and

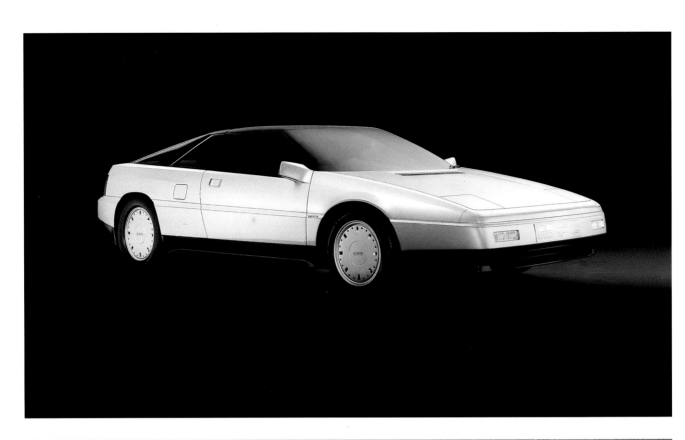

...But this one did – ETNA. Designed by Giugiaro, it was powered by a Siamesed version of the Lotus slant four engine, the four-litre V8 weighing just 414 lbs fully dressed was expected to produce 330bhp and give the car a projected 180+mph top speed with zero to 60 in just 4.3 secs. ETNA was intended as a Lotus Engineering showcase with an advanced VARI monocoque, Active Suspension, photo sensitised glass, proximity radar to warn of impending collisions, a central computer controlling the engine management system, the suspension, air-conditioning and the ABS, as well as a microprocessor managed CVT system as standard. Unfortunately, ETNA remains a stillborn project as Lotus couldn't attract any partners for the costly technology involved. Only the 4-litre V8 saw light of day as it became the basis from which Lotus developed Chevrolet's 5.7-litre LT5 engine used in the potent Corvette ZR-1

ergonomic with VDO instruments clustered directly in front of the driver – and viewable to all but the tallest – with Maestro-derived push buttons controlling minor functions on the outer edges of the binnacle. As with the original Esprit there was precious little space for storing anything inside the cabin, apart from a glove-sized glove box and a small pocket on the rear bulkhead.

The chassis carried over virtually unchanged, though by the time it had been attached to the new body, torsional rigidity was up by a sizeable 22 per cent. Consequently the combination of softer springs and greater vehicle weight – at 2,793 lbs the Turbo had put on 147lbs over its predecessor – meant ride comfort was improved without compromising handling.

On paper the suspension was unchanged: the front consisting of an upper, shorter, wishbone made from a pair of channel-section steel

Above
The interior changed very little over the years, apart from the use of Toyota-sourced switchgear. (I Adcock, Car courtesy of London Lotus Centre)

Left
As the years went by more time, money and effort was spent on developing the Esprit and the Elan, to the detriment of the Excel. Yet it still provided one of the best handling and nimblest 2+2s that money could buy. (I Adcock, Car courtesy of London Lotus Centre).

The Excel never failed to look good on the move, with its clean aerodynamic lines. (I Adcock, Car courtesy of London Lotus Centre)

pressings with the lower one made up of a box section transverse link and a tubular steel arm. The front anti-roll bar was mounted from the outer ends of the lower wishbone.

At the back, the alloy hub carrier is located by a steel tube transverse link at the top and a longer tube at the bottom with a box-section semi-trailing arm pivoted from the leading edge of the hub carrier forward to the chassis' backbone. As if to show there were no hard feelings, Toyota supplied the ventilated 10.1ins front discs and the 10.8ins solid rear ones. Despite the latter now being mounted, outboard unsprung weight was actually reduced thanks to new, lightweight, Lotus-designed alloy wheels manufactured by Italy's OZ. A Goodyear 175/70 SR 14 space saver tyre was stored under the front bonnet as the Esprit had different sized wheel/tyre combinations: 7JKx15ins with Goodyear Eagle NCT 195/60 VR 15s at the front and 8JKx15ins with 235/ VR 15s for the rear.

The new Esprit's powertrain came in for some revision; the engine produced an additional 5bhp in turbo guise and the Renault GTA/25 Turbo five-speed gearbox – with suitable ratios – replaced the Citroën SM unit in all markets except the States which soldiered on with the

During 1983 the Excel's styling was further enhanced. for the SE models. Although the bottom half remained virtually the same – except that the front and rear bumper mouldings were body-coloured – the upper half grew blisters over the wheelarches and a rear spoiler was tacked onto the boot further reducing rearward vision. (Focalpoint)

older transmission for some time to come. But did all these changes amount to a better car? According to the December '87 issue of *Car*, the answer was an emphatically positive: "…the Esprit represents a sizeable leap forward by Lotus. It's a more mature car, a much more balanced car. It will do Lotus' reputation as a serious rival to Porsche and Ferrari a powerful lot of good."

Moreover, when the same magazine put an Esprit through its performance paces a year later, 0-60mph was banished within 5.3secs and a top speed approaching 158mph attained. These impressive Porsche and Ferrari equalling figures were tempered by some doubts gathered after driving the Esprit for 5,000 miles in a week: "The Esprit is not the best supercar, it lacks two crucial ingredients. Its engine…has nothing like the flexibility, the throttle response, or the music of the Porsche 911 or Ferrari GTB. It also lacks the external beauty: its body is not as well crafted…nor as distinctively styled."

Meanwhile the Excel soldiered on, getting a minor facelift and further chassis tweaks. The kerb-prone front spoiler had a flexible rubber lip added to its lower edge, the rear spoiler was redesigned to make

reversing easier and spats were added to the rear wheel arches. As for the chassis, the suspension geometry was altered to maximise new spring and damper rates and the 215/50ZR15 Goodyear Eagle tyres were fitted to new seven-inch alloys. The engine remained untouched, apart from new hydraulic mounts, a sure indication that the Excel's time was running out as the launch of the new Elan – code-named M100 – drew ever nearer.

To celebrate 40 years of Lotus a limited edition Esprit Turbo was announced at the 1988 Birmingham Motor Show with white pearlescent bodywork – complete with rear-mounted spoiler and a deeper front bib – a two-tone blue leather with matching suede interior and a sophisticated Sony hi-fi system with CD player; this Esprit proved an immediate success with UK buyers.

More significantly, the traditional ACBC badge was given a mild

Twelve months later the Excel also received a minor update with a new bonnet, smaller rear wing than the SE's and different wheels

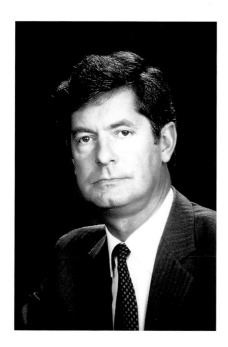

By 1985 Lotus was on a surer financial footing, but the need to grow and develop new models required major financial investment – money that its current owners were unable, or weren't prepared, to spend. The hunt for a buyer was on and by January 1986, GM had been courted and convinced to purchase Lotus. Out went Wickens and all the other shareholders, Toyota were the last to give up their interest, and Michael Kimberley – who had been drafted into Lotus by Colin Chapman in 1969 to develop the Europa – was eventually appointed managing director and chief executive officer

update with stronger colouring and letters; nevertheless, it was instantly recognisable as the true Lotus crest.

In May 1989 the SE version of the Turbo was launched. With 264bhp on tap from its 2.2-litre engine it would rocket to 60mph in 4.7secs and 100mph in 11.9secs before attaining a top speed of 163mph. For once, these figures were modest as *Fast Lane* proved when it tested the car, achieving 4.5secs for the zero to 60mph dash. However, the Esprit was much more than a phenomenally quick car; it was powered by a 'green' engine which complied to the Californian exhaust regulations, the world's toughest. Lotus had also been stung by criticism of the Esprit's chassis, ride and handling and the general lack of build quality. The Esprit Turbo SE was meant to address all these shortcomings.

The front suspension geometry was revised to eliminate anti-dive characteristics, pitching movements and improve ride. Lighter steering effort, especially at low speed, was achieved by reducing caster angle from 2.5 to one degree and increased bump travel with revised bump-steer characteristics further enhanced ride and straight line stability.

Stiffer front springs were fitted and, although the rears remained unchanged from the '89 Esprit, gas-filled twin-tube dampers were used at the back for quicker response.

Goodyear worked closely with Lotus to specially develop new Eagle ZR tyres for the SE: 215/50 ZR 15s mounted on 7JK x 15ins rims at the front and 245/50 ZR 16s on wider 8.5J x 16ins for the rear. The 50-series low-profile tyres featured a new tread compound while the rears had a unique purpose-designed carcass construction.

The additional rear wing and front spoiler were not just cosmetics, but served a real function in giving the car neutral balance throughout its speed range. Front and rear lift was tuned to only a few points above zero and the Peter Stevens-led design team paid particular attention to cross-wind stability so as to achieve a linear rate of change.

In addition to the central door locking and electrically-operated fuel filler caps which were introduced on the 1989 specification Esprit Turbo, the SE also featured a burr Elm fascia, full leather interior trim, removable glass sunroof and air-conditioning. An ice-warning light which came on once the ambient temperature dropped below five degrees Centigrade was fitted for the first time.

But was the £42,500 Esprit worth it? According to *Car* magazine it was: "Fact is, the only cars more accelerative [than the SE] are limited-edition, telephone-number-expensive monsters such as the F40 and 959. The SE Esprit has the others comfortably covered. In its newly fettled SE guise, there's no doubt that the Lotus is Britain's best supercar. It's not just Britain's best: it's one of the world's best, and without doubt the world's best value."

Above
Still, there was one thing that few people could argue about and that was the Esprit's handling; in fact, the non-turbo car was, in many ways, a more benign vehicle than its more powerful brother. (Focalpoint)

Above right
An overhead shot of this 1984 Turbo Esprit shows just how limited rear vision was with the venetian blind slats covering the rear window. (Focalpoint)

Right
The Esprit's interior was still pretty lavish with its all-leather trim. However, it was coming under increasing press criticism as it was very uncomfortable for any tall person to drive; note the handbrake's inconvenient location. (Focalpoint)

Above

Automatic versions of the Excel were dubbed SA and first appeared in 1986 with a ZF four-speed auto and the 180Bhp engine; the bumper inserts had appeared earlier together with the new integral front bib and air intake. (Focalpoint)

Left

If ETNA was ultimately nothing but a dream then, at least, Lotus could take some solace in the new Esprit. Penned by Peter Stevens, who had already been successful in updating the Eclat into the Excel, it retained the essence of Giugiaro's original at the same time bringing it up to date. This styling model dates from 1987

Above
By 1986, when this picture was taken, Giugiaro's design was beginning to look a little dated and although build quality had improved it still wasn't on a par with its main rivals. (Focalpoint)

Above left
For those that didn't want the Turbo's searing performance there was always the normally-aspirated version; this is HC model form 1986-7 with its High Compression 172bhp engine. (Focalpoint)

Below left
To maintain sales impetus and interest in the Turbo Esprit, Lotus introduced an HC version with an additional 5bhp and 20lbs/ft of torque, though the latter came in at 250rpm higher than in the old car. (Focalpoint)

Lotus was on a roll. The Esprit had thrust the company into the top ranks of supercar producers and any fears that its independence might have been compromised by GM's big-brother attitude were finally quashed. The GM relationship was, seemingly, harmonious; the SE's sophisticated engine management system was only available because of Lotus' link with GM.

As the summer months passed there was growing excitement at Hethel, and anticipation amongst the specialist press, as everyone knew that M100 – the eagerly-awaited new Lotus Elan – was about to appear.

If you're wondering where you've seen those rear lights before, they're originally from the Rover 3500. (Focalpoint)

Esprits and Eclats shared the same production facilities at Hethel and were built alongside each other. (Focalpoint)

Above left

Turbocharged models, now known more logically as Esprit Turbos, featured the glass back which did nothing for headlamp reflections in the rear view mirror when driving at night. This is a post-1988 car as can be seen from the Citroën CX door mirrors – what would supercar producers do without them?

Left

An American specification Esprit Turbo with repeater flashers and, yes, the wheels are supposed to be like that. (Focalpoint)

Above

The fascia still retained an element of Giugiaro's boomerang design, but had now evolved into more of a pod; minor controls and switchgear are still courtesy of more mundane cars. (Focalpoint)

Above
The principal difference between the UK special and that sold in the States, apart from being right-hand drive, were the rear wheels – not dished for the home market

Right
Esprit production in full swing at Hethel. (Focalpoint)

Above

The excuse for this car was to celebrate "Forty years of achievement" and each car had a numbered plaque mounted on the fascia to record that. (Focalpoint)

Right

A blend of blue leather and suede was used to trim the special's interior and veneer made its first appearance in a modern Esprit

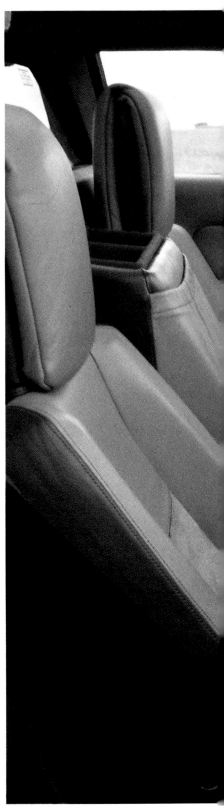

Above
Building Esprit chassis, this one is destined for an SE. The Eagle chassis was specifically tuned to work in conjunction with the similarly named tyres Goodyear developed for the SE model. (Focalpoint)

Above right
Body trim is adjacent to the assembly lines at Hethel, so the seats etc. can be transferred virtually directly into the cars. (Focalpoint)

Right
Esprit Turbo SE's roll down Hethel's production line, which hasn't changed that much from the days when the original Esprit was produced

Above

Much of Lotus' income came from its Engineering consultancy and projects like SID – Suspension, Isolation and Dynamics. This re-bodied Esprit had a Kevlar/Carbon Fibre/GRP monocoque and a Metro 6R4 powertrain. It was developed to help Lotus, and its clients, understand and develop the dynamic interaction between the car's various components

Right

In mid-1989 Lotus unleashed the Esprit Turbo SE with 264Bhp on tap, revised chassis settings, a 163mph top speed and 0-60mph in 4.7secs and all for less than £50,000. Surely the bargain supercar of the decade. Here it's pictured outside Ketteringham Hall, the home of Team Lotus

Above left
Tan leather and matching carpets blended well with the veneer fascia. (CTP)

Left
This overhead shot gives a good impression of just how wide an Esprit is, at 73.2 ins care is needed threading the car through narrow lanes and thick traffic. (CTP)

Above
In the late Eighties Hethel underwent a massive building programme with new facilities being put in for the Elan (the two yellow roofed building in the centre and to the right), while the building work which can just be seen on the left of the picture is the semi-anechoic chamber under construction. The test track can be seen sweeping across the centre of the photograph

Elan to Bugatti

A new Elan had been Kimberley's dream since 1976, but for a variety of reasons it had never been realised: Chapman wasn't keen on the strategy as he wanted to take Lotus further upmarket into the four door, four-seater sector. Then Lotus was hit with the oil and financial crises which almost finished the company off.

However, 1981's agreement with Toyota meant that Lotus had access to one of the world's largest parts bins so a new powertrain and sub-assemblies like suspension, steering etc. would be relatively cheap; consequently, M90 was born.

Penned by Oliver Winterbottom it's a neat, wedgy design in the same fashion as the TVR Tasmin and original Toyota MR2, but nothing sensational or innovative as its illustrious forebear had been – apart from a two-piece hood which allowed you to remove the centre section, but keep the rear window in place, hardly earth-shattering. It followed the classic Lotus concept of a backbone chassis with front engine/rear-wheel drive, but no one at Hethel could get really excited about it. Chapman's death and the uncertainty that followed meant M90 was put on the back-burner until November 1983 when it was revived as X100.

However, the new car bore no resemblance to its immediate forerunner apart from using Toyota componentry. For some time Lotus' ride and handling department – headed by Roger Becker and with ex-Lotus F1 driver, John Miles heavily involved – had been working on a number of front-wheel drive projects for third party clients and Miles, in particular, was convinced that Lotus should go front-wheel drive for its new Elan. There were other reasons though: most of the world's majors were going front-wheel drive and any rear-wheel drive powertrain package which Lotus bought was likely to be yesterday's technology. Lotus needed to be at the forefront, moreover Toyota would probably be the supplier involved and all their small and medium cars were going front-wheel drive.

Peter Stevens had taken over the design role and his brief was to produce a coupe and convertible which owed a family resemblance to the Etna supercar. But try as he might, he and Lotus' director of design, Colin Spooner just couldn't make X100 work. For two years the

*When GM bought Lotus, the Elan project was revived and the styling put out to tender: Giugiaro's vision of an Elan is a neat coupe that owes too much to the likes of Isuzu's Piazza, while GM's Chuck Jordan produced this concept which would have been fine for a Pontiac (**left**), whereas Peter Steven's bold approach was just the sort of visual challenge Lotus was looking for. Here (**above**), it's being reviewed by senior Lotus personnel including Mike Kimberley (centre, holding the door), chairman Alan Curtis (to Kimberly's right) and Peter Wright, Lotus' technical director and the father of Active Suspension, by the car's nose*

designers struggled to make X100 better; lowering the nose, slimming it down and generally taking the bulk out of the car's shape, but to no avail. In December 1985, work on X100 was put on hold as the build-up to GM's takeover gathered momentum, even so Lotus' management professed themselves happy with the car's front-wheel drive contents – it was just the packaging which wasn't right.

In March 1986, only weeks after GM's takeover, Kimberley convened a meeting at Hethel to announce that M100 – Lotus had reverted back to the Chapman style of project numbering with an 'M' rather than Wickens' 'X' ratings – was a viable project, but that its design was now open to three competitors: Giugiaro's ItalDesign, GM's design centre headed by Chuck Jordan and Lotus' in-house team of Peter Stevens and Simon Cox lead by Colin Spooner.

Eight months later Spooner's team made its presentation to the Lotus board and unanimously won approval for their concept over the full-size styling clays offered by ItalDesign and Jordan. While ItalDesign's reactions to losing are not on record, Chuck Jordan later vilified Stevens' efforts in front of the British press even before the Elan appeared.

With GM on the scene, it was unthinkable that the Americans would allow a Toyota powerplant to be used, or that the Japanese would supply one. Indeed, that turned out to be the case and Lotus had to trawl GM's

Above left
One of the many endurance vehicles Lotus ran during the Elan's development, heavily disguised with black tape and false sills

Below left
Mike Kimberley unveils the Lotus Elan at the 1989 London Motor Fair for its world debut

global parts bin for a replacement power unit. That search eventually finished in Japan at Isuzu who had a new 1.6 litre four-cylinder under development. The fact that the engine had yet to be signed off, meant Lotus could influence its design and were able to persuade the Japanese to reshape the plenum chamber to make the engine more compact as well as assisting in helping programme the engine management system and to satisfactorily tune the Noise, Vibration and Harshness characteristics. The main advantage of the Isuzu power unit was its compactness compared to the Toyota which meant Stevens could repackage the entire car.

When the new Elan was finally revealed at London's 1989 Motor Fair it drew praise and criticism for its styling in equal parts – it was a design people either loved or hated, there was no in between. From its sharply drooping bonnet, through the complex waisting – virtually giving the car an hourglass figure in plan view – to the abruptly chopped-off boot with practically no overhang it was a daring assault on the conventional designs which bounded open two-seaters. It boasted careful aerodynamic tuning: 0.34 Cd with the hood in place and 0.38 with it down, there was negative lift front and rear and much detailed wind tunnel testing undertaken to ensure the car remained stable under all conditions.

Beneath the complex bodyshape there was a revolutionary chassis and suspension system: Gone was the usual Lotus backbone-type chassis; in its place a one- piece three mm VARI moulding riveted and bonded to steel reinforcing outriggers was used which, in turn, was bolted to an octagonal backbone chassis. Complicated it might have been, but its 6,600 lbs ft/degree of torsional stiffness was high for a conventional hatchback, never mind an open two-seater. However, the real revolution which separated this Elan from other front-wheel drive cars was the patented 'Interactive wishbone' front suspension designed by Lotus engineers John Miles and Jerry Booen with input from Roger Becker.

Both front suspension assemblies are mounted on aluminium-alloy rafts which, in turn, are bolted to the chassis at three mounting points; two of these – those that control vertical and lateral movement are very stiff – while the third, softer mounting, restricts longitudinal compliance and reduces road noise.

Conventional wishbone systems mounted directly to the chassis employ soft bushings to eliminate road noise and promote a comfortable ride, but they also result in suspension changes – especially castor – during driving. What Lotus did was to mount the wishbones as stiffly as possible to the raft using metal and rubber bushing rather than solid rubber, thus maintaining the suspension's geometry under cornering, braking and acceleration.

A 22mm tubular anti-roll bar was fitted at the front and a solid 14mm

one at the rear. The independent rear suspension with its upper link and wide-based lower wishbone was similar to that of the Excel, but adapted to the Elan's larger production run and different packaging requirements.

Turbo models benefited from power-assisted steering – 2.9 turns lock-to-lock – while the normally-aspirated cars had manual steering with 3.1 turns lock-to-lock. If the raft system was a breakthrough in suspension design, then the manufacturing technology used in the body assembly was equally daring.

In a complete break from its traditional double 'bath tub' arrangement Lotus used for the Excel and Esprit, the Elan's bodywork was built up of a number of smaller separate panels to allow future design freedom. The panels being jig assembled for accurate and consistent control of the assembly.

Lotus Engineering developed, and patented, the 'Fibreform' process – a method by which pre-formed fibre panels could be self-located in

Above
The first Lotus Elan SE, appropriately in British Racing Green, was bought by the Patrick Motor Collection in Birmingham as part of their permanent display of cars. (Lionel Photo)

Right
All the early Elans were painted in Silver, here an Esprit Turbo SE and an Excel flank the new Lotus with Ketteringham Hall providing the backdrop. (Monitor)

Nickel-plated, heated moulds. This system had a number of inherent benefits over the original VARI process in that ambient temperatures didn't affect curing times, the finish was consistently better and of higher quality, the tooling lasted longer and it was a much faster production process – essential as Lotus planned on building 3000 Elans a year.

While the body production and suspension technologies were typical products of Lotus Engineering foresight, the decision to use a Japanese engine was controversial. Isuzu, partly owned by GM, was one of the minnows in the American conglomerate's empire and needed a boost to its image – the link with Lotus was ideal. Lotus had already done some ride and handling work on the Isuzu Piazza coupe, so the link was well established and, in fairness, the iron block/alloy head power unit jointly developed by Lotus and Isuzu was one of the best in its class. Two versions were offered, though the normally-aspirated model only ever appeared on the price list and very, very few were actually produced, both based on the same 1,588cc, 16-valve, dohc block: the 165bhp, 148 lbs/ft turbo model and the 130bhp, 105 lbs ft fuel-injected version. Compression ratios were 8.2:1 for the blown engine and 10:1 for the less powerful unit.

A five-speed gearbox with common ratios was employed, though the turbo had a higher – 3.833:1 – final drive compared to the normally-

aspirated model's 4.117:1. Top speeds were 137 and 122 mph, respectively, while the more powerful car could hit 60mph from standstill in 6.7secs, 9/10ths of a second faster than the 130bhp model.

Simon Cox was charged with designing an interior which, with the car's hood down, had to be as stylish and professional as the rest of the car – no more stretching leather over plywood forms as had been Lotus' past practice.

It helped that Cox, too, could raid GM's parts bins for switches, air vents and instrumentation, but even so, they were cleverly disguised by using red symbols and figures rather than conventional white. The facia moulding gave a true cockpit feeling to the driver and the entire interior had an integrated feel to it that no previous Lotus had possessed.

The new Lotus Elan was the affirmation of GM's faith in the marque; it had swallowed up £35 million in new production and paint facilities – as well as design and development costs – at Hethel. This was the car, which, in Kimberleys' words, returned Lotus "to the sort of volumes we used to produce from this factory."

Above
Elan's waiting final inspection at the end of the newly expanded production facility

Left
In its element, high-speed cross-country dashes with the roof down, the Elan set new standards in front-wheel drive handling and stability. (I. Adcock)

His and GM's faith seemed justified once the world's press got their hands on the car. *Autocar & Motor* described it as the "The first front-drive supercar" and urged the reader to: "sell your grandmother and the rest of your family, your dog, your cat and anything else to hand to have the new Elan."

While *Car*'s Gavin Green spoke for all the press – many sceptical that a real sports car, especially a Lotus, could have front-wheel drive and still be fun – when he wrote: "With the Lotus Elan, front-drive reaches a new plane of excellence. This car quashes the serious driver's prejudices against the puller powertrain. The Elan is a turning point in car development…Down a winding road, the little Lotus strings together corners with a greater fluency than any sports car I know. Its smallness, its nimbleness, is a huge boon. You can always hit the apex, accurately use all the road. Even on slightly greasy surfaces, there is an accuracy, a sure-footedness, that no rear-drive car would have imbued."

Green concluded that "It's a winner, the new Elan. One of the most significant sports cars of all, and one of the greatest." Fulsome praise, indeed, and typical of that written by the press. The hype was translated

Above
The magnificently brutal Lotus Carlton, or Lotus Omega if you happened to buy a left-hooker. Previewed at the 1989 Geneva Salon, its proposed specifications – 350+bhp and 170+mph top speed – stunned the motoring press. (Paul Debois)

Right
Harnessing the power and torque from the twin-turboed engine was no mean feat, but Lotus' ride and handling specialists excelled themselves to produce a car with a boulevard ride that could be hurled round corners like a nimble two-seater. (Paul Debois)

into serious enquiries, so that by the end of the 1989 London Motorfair – and the Tokyo Motor Show which ran virtually simultaneously – Lotus had serious sales enquiries for more than 1,000 Elans. On the surface it looked as if Lotus, thanks to a brave new design, had left behind, forever, the years of uncertainty and doubt which had dogged the company. It appeared as if Lotus could do no wrong for earlier the same year it had stunned the motoring world with the Lotus Carlton/Omega which was a last minute debutante on GM's stand at the 1989 Geneva Salon.

Bob Eaton, president of General Motors Europe and longtime Lotus fan, was determined that Vauxhall and Opel should shed their blue-collar image and add a bit of bravado to the product ranges. What better way of doing that, then, by getting Lotus to tweak the top-line Vauxhall Carlton/Opel Omega.

Lotus' concept was audacious, they simply wanted to produce the

It is doubtful that anyone will repeat an exercise like the Lotus Carlton again. A magnificent folly or the last of dying breed? Probably both. (GME)

Powerhouse! Each engine was handbuilt on a special line at Lotus; months of careful development had turned Opel's standard 24-valve straight-six producing 204bhp and 199 lbs ft torque, into a fire-breathing monster pumping out 377bhp and a massive 419 lbs/ft. (Paul Debois)

world's fastest four-door saloon. And they achieved it. Its specifications read like competition car credentials, rather than a limited edition, high-speed luxury motorway express: The 3.6-litre, twin-turboed straight-six engine developed 377bhp at 5,200rpm and 419 lbs/ft torque at 4,200rpm. Mated to the same six-speed ZF gearbox used in the Corvette ZR1, this gave the Carlton/Omega a top speed of 177mph and 0-60mph in 5.1 seconds. Its in-gear performance figures were equally impressive and it became the first car *Autocar & Motor* tested to achieve a 140-160mph in-gear time (14.2 secs for the record).

Revised aerodynamics developed by Lotus ensured the car remained stable throughout its speed range without increasing the standard car's 0.31 Cd. The interior, meanwhile, was swathed in Connolly leather and walnut door cappings.

The whole effect was capped off with a quartet of monstrous tyres

and wheels: 17x8.5 ins front rims with 235/45ZR17 tyres and larger, 9.5ins rims at the back with 265/40ZR17 rubber.

Officially the car transmuted from being an Opel/Vauxhall into a Lotus, as it was entirely rebuilt at Hethel and homologated as a Lotus. However, the build sequence was tortuous: completed cars were shipped from GM's Russelheim factory and then stripped out, the engines, transmissions and any other salvageable items returned to Germany and the car gutted and completely rebuilt at Hethel.

Considering its performance potential and the build programme's complexity, the £48,000 price tag seemed good value for money. However, unlike the Elan, the Carlton was not greeted with universal praise by certain sections of the media. During the slack summer of 1990, *Autocar & Motor* lambasted it advocating that GM should limit the Lotus Carlton's performance, intimating that since the car was merely a 'Vauxhall' it was incapable of coping with its performance potential. This neatly ignored the wholesale changes which Lotus engineers had made to the car's ride, handling and braking systems. Apart from its basic shape the Lotus Carlton bore about as much resemblance to the Vauxhall version as Linford Christie does to a first-time jogger.

However, once the media got to play with the Lotus Carlton, attitudes changed: "The miracle of the world's fastest production saloon is that it possesses a chassis that tames the fury so effectively, you don't need to be Schumacher to drive it well. Be in no doubt, the LC is stupendously rapid…and such is its overtaking ability and braking power, it's incredibly safe." The magazine? *Autocar & Motor*, November 1990.

Meanwhile, Hethel was struggling with the realities of getting the Elan into production. Promises made in October 1989 that the car would be on sale "within months" started to look empty and it wasn't until the following April that the first few customer cars started to trickle out of the factory. Production problems with the new VARI system, the pop-up headlamps and the hood dogged deliveries. Even so, pent-up demand for the Elan – which went on to win the 1991 Design Council Award for motor industry products – ensured there were still plenty of customers prepared to wait. By the end of 1990 over 1200 Elans had been delivered, but that only tells part of the story.

The Elan facility had been designed to eventually produce 3000 cars a year and Lotus' marketing strategy was that one-third would be sold in the USA, a third in UK/Europe and the balance in other markets (especially Japan).

While UK sales got off to a reasonable start, the all-important US market was way behind schedule, mainly due to increased development time to get the Elan's air-bag engineered. The lateness of the car's appearance was further compounded by its $40,000 price tag which pitched it right up with Corvettes, Nissan 300ZXs etc, and way beyond the $30,000 tag originally forecast.

Another profitable market which Lotus never satisfactorily penetrated was Japan. As part of the engine deal, Isuzu was to market the Elan in Japan, unfortunately they saw it as an Mazda MX5 rival and wanted to price it as such whereas Lotus wanted to pitch it against the more costly Porsche 944 cabriolet. The impasse was never resolved and Lotus lost out on yet another potential profit earner. By the end of 1991 Lotus had built a further 2060 Elans, still a third short of the production facilities potential. To compound matters, the bottom had dropped out of the super sports car market and just 125 Esprits had been built.

In an effort to revitalise the ageing supercar, the Esprit was repackaged for 1992. After years of customers and the press complaining about the Esprit's cramped interior and narrow-opening doors, Lotus redressed the problem.

Some clever redesigning resulted in seats over two inches wider, a slimmer central tunnel gave more elbow and passenger space, a re-designed bulkhead added 1.4ins to the cockpit length and a new pedalbox increased foot-well space. A new hingeing system saw the doors open a further 15 degrees so passengers and drivers could at least get in and out of the car without performing circus-like contortions.

Above
Adrian Palmer, who took over as managing director from Mike Kimberley in 1991. Two years later GM sold Lotus to Bugatti. Only time will tell if these two companies can survive

Above right
Lotus added a hardtop option to the Elan range in October 1991 but it never went into production

Right
When GM pulled the plug on the Elan in mid 1992 after just 3,857 had been built, it was believed that an era had come to an end. Little did any one dream that two years later it would be back in production. (I. Adcock. Car courtesy of Mike & Gill Bishop)

Left
Many car manufacturers developed from humbler origins as bicycle makers, but not Lotus who reversed that trend and in 1992 produced the Superbike which took Chris Boardman to a gold medal in that year's Olympics

Above right
In action in the States, the Esprit racer proved more than a match for the likes of Nissan, Chevrolet and Porsche, although this wasn't always translated into "Win on Sunday, sell on Monday"

Below right
Fortunately Lotus had the good sense to realise that that wing wasn't making the marque any new friends and in early 1993 launched the Lotus Esprit S4 as the only version of the mid-engined two-seater offered. The very subtle, but extensive re-styling has freshened up the car's looks and marginally improved leg, head and shoulder room for driver and passenger. (I. Adcock)

Externally, the car came in for a minor facelift with the deletion of the glassback rear window and, for the SE only, a new rubber lip on the front bib and revised engine bay flow. Most controversially of all there was a high-mounted rear wing to compensate for deleting the old smaller spoiler and the sloping rear window. It might have increased the car's top speed to 165mph and reduced its Cd to 0.33 and the whole aerodynamic package might have resulted in virtually zero lift front and rear, but that wing was possibly the ugliest appendage ever put on a car and totally out of place.

Just the chargecooled SE model was offered with the device, the only other Esprit now on sale being the 215bhp Esprit. At the same time Lotus announced a hardtop for the Elan. Featuring a double-skinned composite construction, a heated rear window and an interior courtesy light it was scheduled to go on sale the following year, but none were ever sold to the public.

Once again it looked as if Lotus had snatched defeat from the jaws of victory. A global recession, luxury taxes imposed on imported cars in the USA, a failure to secure a distribution deal in Japan and, above all, the Elan was an impossibly complicated car to build that was always going to be marginal on profits conspired against Lotus. But such worries shouldn't have concerned Lotus, tucked safely as it was in the bosom of GM. What were a few losses to the world's mightiest car corporation? The amount of money Lotus was losing must have been the equivalent of gnat's blood compared to the amount of red ink that must have filled

GM's accounts; in 1991 it made a vast $4.5 billion loss, at the time the biggest in corporate American history.

Apparently it did concern Lotus and once Bob Eaton – Lotus' fairy godfather within GM for so many years – had unexpectedly quit in March '92 to become chairman of arch-rivals, Chrysler. The accountants' writing was on the wall for Lotus and it didn't take long for the Detroit beancounters to make a decision.

At the beginning of March, Adrian Palmer – Lotus' new MD in the wake of Kimberley's departure to new GM pastures in the Far East – announced that all car production at Hethel would stop for five weeks and that the Excel, now in its eighteenth year, would be dropped altogether.

However, on Monday June 15 1992 the inevitable happened: the Elan was killed off, just 30 months and 3,857 cars after it had been launched. The signs had been posted earlier in the year when the decision was

Above
The interior, though finished in sombre grey is roomier; GM and Rover sourced switchgear has replaced the out-dated componentry that was once used. (I. Adcock)

Right
In profile, the Esprit is little changed save for a deeper front bib, air ducts for the front brakes, scalloped air scoops cut into the sills, and twin wings at the back – the flying bridge and ducktail lip obviate the need for the glassback hatch. (I. Adcock)

taken to cut back Elan production to 2000, but even that and a midnight oil project to take cost out of the Elan failed to save one of the world's greatest sports cars. Matters could only get worse for Lotus, and did, when GM announced that just 950 Lotus Carlton/Omegas would be built – 150 short of the predicted run – and that production would finish at the end of 1992.

So, in 12 months, Lotus had gone from dizzy heights to the dismal depths of despair and Hethel, once buzzing with new jobs and unbridled enthusiasm, was more like a ghost town with two dormant production facilities, the Elans and the Lotus Carlton/Omegas.

It would be difficult to imagine matters getting much worse at Hethel, but they did as 1992's Esprit production was a meagre 173 and none were built for the US market. The questions that lingered on everyone's lips were simple: could Lotus survive; moreover did GM *want* Lotus to survive?

Above

Extensive use of lightweight composites to help reduce body weight was a feature of the Sport 300 as were ABS brakes and power steering. (I. Adcock)

Left

As well as the wheel arch extension the competition programme created revised aerodynamics with a deeper front bib and a more elegant solution to the rear wing problem than that used on the S4. (I. Adcock)

Doom and gloom might have been the prevailing atmosphere at Hethel, but they were putting on a brave face. At the 1992 Birmingham Motor Show, Lotus displayed a concept version of the Esprit. Based on the Esprit X180R race version which had been enjoying considerable success in the American Supercar Championship with Doc Bundy winning the Driver's title, the Sport 300 featured a tuned version of the 2.2- litre engine producing 300bhp achieved by reworking and porting the cylinder head, using a hybrid Garrett T4 turbocharger, larger injectors, a bigger chargecooler and linking the airconditioning radiator into the chargecooling system to increase its efficiency. Extensive use of composites shaved 250lbs off the bodyshell, complete with wheelarch extensions and – thankfully – a neater rear wing arrangement. The chassis was stiffened, bigger – 13ins front and 11ins rear – brakes used, gorgeous OZ three-piece alloy wheels (8.5x16ins up front and bigger, 10.5x17ins at the back), specified and shod with equally impressive 245/45x16 and 315/35x17 Goodyear GS-C tyres. While the suspension was heavily revamped with new geometry settings to give more grip and balance as well as modifications to the top links, bushings, damper and spring rates.

At its launch there was vague talk of production, but nothing concrete. In fact most people wondered if Lotus would survive into the following year. The question was partly answered in February 1993 when the Esprit S4 was launched. From a distance it might have looked like its forebear – thankfully with a neater flying wing half way up the buttresses – but close up it was a very different car.

Sitting on elegant five-spoke 17ins rims girded by asymmetric Goodyear Eagle GS-A tyres (215/40 on 7J rims at the front and 245/45 on 8.5J rims at the back), the new Esprit boasted a revised front spoiler which recalled the original Elan's air intake, new side skirts, door skins, front and rear bumpers and subtle changes to the 'A' pillar, engine vents and bonnet. At last, the Morris Marina door handles which had blighted the Esprit since its birth were abandoned in favour of flush Rover 200-sourced body-coloured units.

Once inside, there was evidence of more GM componentry: electric window and mirror controls, door handles and stalk controls; it makes you wonder why it took Lotus, and GM, so long to dump the Esprit's old switch gear which always detracted from the car's best features. Carbonfibre now decorated the facia in place of wood veneer and a fresh, chunky Nardi steering wheel topped a new steering column.

However, what the steering column was attached to was important. For the first time Lotus fitted power steering to its mid-engined supercar. Gone forever were the muscle-building efforts required to park an Esprit. It took Lotus two years to develop the Saginaw system for

Above right
Massive split OZ alloy wheels clad with Goodyear GS-C rubber wear forced Lotus to fit the make-shift looking wheel arch extensions. (I. Adcock)

Below right
Re-trimmed interior sports Alcantara trim, competition-style bucket seats and drilled pedals. (I. Adcock)

Esprit and it demanded subtle changes to the front suspension geometry as well as a new anti-roll bar, springs and dampers to work effectively.

The 2.2-litre engine maintained its 264bhp output, but all-new head, block and sump castings increased stiffness, a more durable cylinder head gasket was employed and the engine management system re-mapped to give more torque at lower revs. A revised high-torque version of the Renault gearbox was also used.

At £5 short of £47,000, the Esprit was getting expensive – but were all these changes worth it? In *Car*, Richard Bremner gave the S4 a qualified thumbs up when he wrote: "If the Lotus were that bit more refined, had less jerky, more euphonious driveline (and that means adding cylinders), less alarmingly arranged pedals, and quality that didn't feel like it was from the cheaper end of the street, I might be swayed."

Still, it was going to be a rare car; Lotus announced that only 300 Esprit S4s would be made in 1993 plus a further 50 Sport 300s. Surely, not enough to keep Hethel running as a viable producer? Ten days after the S4's launch – 24th February – the news leaked out that Lotus was up for sale. GM tired of supporting the company, and with a new management team at its head determined to bring the American giant back into profit, were off. What everyone had known would happen when the Elan was killed off, but had been afraid to admit, had become reality: Lotus was back on its own.

Managing director, Adrian Palmer tried manfully to put a management buy-out together, but it was never really on. City investors had been burnt too often in the past with specialist car producers and in 1993, when the market for luxury sports cars was at its lowest, there was little faith in Lotus. A chequered history, which had seen Lotus stagger from insolvency to crises and back again, can't have helped Palmer's cause.

GM had, at least, performed one honourable act and that was to right off Lotus' £54 million debt, so any new owner would have a clean balance sheet – but they would need tens of millions to finance a new product range. As Palmer and his team struggled to get the financing, things deteriorated further when 70 employees were made redundant in early summer 1993 as orders for cars and, crucially, consultancy work for Lotus Engineering started to peter out. Even Touche Ross, Lotus' auditors, were openly questioning the marque's long-term viability when an unlikely saviour appeared on the horizon – Bugatti.

The legendary French marque had been revived in 1990 by Romano Artioli – a wealthy Italian entrepreneur and car enthusiast who counts amongst his business interests the Lotus franchise in Italy – and a group of mystery backers. Artioli had long been an admirer of Lotus Engineering's technology – especially its Active Suspension and Adaptive Noise Control know-how – and ownership of Group Lotus (excluding

Above right
The Elan's cockpit looks much the same as before , bar new seats and a re-profiled central tunnel which frees up much-needed elbow room. (I. Adcock)

Below right
John Miles, who co-developed the original front-wheel drive Elan's raft front suspension, worked with Tony Shute – who honed the Lotus Carlton's handling – to make the S2 more responsive by re-valving the power steering and re-calibrating the Monroe shock absorbers. (I. Adcock)

Team and the Millbrook Proving Ground), would give Bugatti instant access to that technology.

Rumours also abounded that there were bids in for the mothballed Elan production line, including proposals from a number of Far Eastern consortiums and Kia, the South Korean car producer.

However, by the end of August all was settled and Bugatti had bought Group Lotus from GM for an undisclosed sum, rumours maintaining that it was £30 million.

As this is being written Bugatti's plans for Lotus are unknown: there is talk of an Esprit replacement, a contemporary Lotus 7 and, possibly a modern version of the 2+2 Excel. (The editorial office at Osprey know a thing or two, but are sworn to secrecy at the moment.) The only concrete news is the Elan's return. When Lotus ceased producing the front-drive supercar, it held in stock some 800 Isuzu powertrains. That, together with GM's decision to wipe out all of Lotus' debt, meant any new owner could put the car back into production for very little cost, but maximum profitability. Artioli and Palmer confirmed that the 'Mark II' Lotus Elan would be in production in mid-1994, with only minor changes over the original, including minor interior revisions and upgraded suspension with stiffer springs and bushes, bigger 16x7ins wheels with 205/45ZR –16 tyres, while the doors would be updated to incorporate the latest side-impact technology. Its sale price, according to Palmer, should be near the £22,750 tag the original Elan SE carried when it was withdrawn from sale.

So, as this book closes, a new chapter in Lotus' varied history opens. Motoring enthusiasts, especially Lotus owners, will fervently hope that the men and women who are now guardians of the 'ACBC' logo appreciate just what it is they possess, and that they have learned the lessons of history. After all, we all thought Lotus had been saved when David Wickens and company acquired Lotus, then we all knew Lotus had been saved when GM swept in. Perhaps, with Bugatti, it will be third time lucky. I hope so. Though if any company's history indicates that nothing is certain and that white knights sometimes give up if faced with too many dragons, it is the history of Lotus.

Technical Specifications

TYPE 14 LOTUS ELITE

Engine: Coventry Climax Type FWE
Cylinders: 4 in-line
Fuel supply: Single 1.5ins or twin(*) SU carburettors
Cubic capacity (cc): 1,216
Bore/stroke (mm): 76.2/66.7
Compression ratio: 10.0:1
Valve gear: Single OHC
Ignition: Coil and distributor
Main bearings: 3
Max power (Bhp/rpm): 71/6,700 83/6,250*
Max torque (lbs ft/rpm): 76.5/3,500 75/4,750*
Transmission (ratios/mph per 1,000rpm)

Overall gear ratios
1st: 3.67/N.A.
2nd: 2.20/N.A.
3rd: 1.32/N.A.
4th: 1.0/N.A.
Reverse: 3.67
Rear axle: 4.5:1
Wheels & Tyres
Centre lock, wire spoke 4.90-15 ins tyres
Brakes:
F/R: 9.5ins/9.5ins discs
Steering: Rack-and-pinion
Lock-to-lock: 2.5
Turning circle between walls: 37ft 10ins

Measurements (ins)
Length: 148
Width: 59.25
Height: 46.5
Wheelbase: 88.25
Track (F/R): 47/48.25
Ground clearance: 5
Kerb weight (lbs): 1,456

Suspension
Front: Independent, transverse wishbones with coil springs mounted on telescopic dampers, anti-roll bar.
Rear: Independent, Chapman strut type with fixed length drive shafts, coil springs on telescopic dampers with auxiliary springs.

Performance figures (mph/secs)

0-60: 13.2 11.1*
0-100: 42.3 33.8*
1/4-mile (secs): 18.8 17.5*
Max speed (mph): 112.3 117.4*
Average mpg: 34.4 31.8*

NB: All contemporary road tests were of cars fitted with optional 4.22 rear axle and not the standard 4.5 unit.

Production figures
Series I Series II (incl. SE, Super, 95, 100 & 105)
280 749, plus upwards of 30 remaining after production ceased

TYPE 26, 36, 45 LOTUS ELAN

Engine: Lotus-Ford
Cylinders: 4 in line
Fuel supply: Two, twin-choke Weber 40 DCOE (Dellortos optional)
Bore/stroke (mm): 82.55/72.75
Cubic capacity (cc): 1,558
Compression ratio: 9.5:1
Valve gear: Twin overhead camshafts
Ignition: Coil and distributor
Main bearings: 5
Max power (Bhp/rpm): 105/5,500
Max torque (lbs ft/rpm): 108/4,000

Transmission (ratio/mph per 1,000rpm)
1st: 2.50/6.6
2nd: 1.64/10.1
3rd: 1.23/13.4
4th: 1.00/16.5
Reverse: 2.79
Rear axle: 3.90:1
Wheels & Tyres
Pressed steel wheels with 4.5J rims, 145-13 tyres
Brakes
F/R: 9.5ins/10ins discs
Steering: rack and pinion
Turns lock-to-lock: 2.5
Turning circle (Between kerbs): 30ft 9ins
Measurements (ins)
Length: 145.25
Width: 56
Height: 35 (38 with roof up)
Wheelbase: 84
Track (F/R): 46.25/48
Ground clearance: 4.25
Kerb weight (lbs): 1,484

Suspension
Front: Independent, wishbones, coil springs and telescopic dampers
Rear: Independent, Chapman strut, wishbones, coil springs and telescopic dampers

Performance figures (mph/secs)

0-60: 9.0
0-100: 24.1
1/4-mile (secs): 16.5
Max speed (mph): 111.9
Average mpg: 25.5

Production figures (road-going cars only):

Series	1	2	3	4	Sprint
	848	1,205	2,650	2,976-3,000	900-1,353

Series 3SE
To all intents and purposes the same as its forebears except for:
Engine
Fuel supply: Twin Weber 40DCOE18 carburettors
Max power (Bhp/rpm): 115/6,000
Max torque (lbs ft/rpm): 108/4,000
Transmission (ratio/mph per 1,000rpm)
1st: 2.97/N.A.
2nd: 2.01/N.A.
3rd: 1.40/N.A.
4th: 1.0/N.A.
Reverse: 3.32
Final drive: 3.77:1 (3.55:1 as tested by the press)
Performance figures (mph/secs)
0-60: 7.6
0-100: 20.9
1/4-mile (secs): 15.7
Max speed (mph): 122
Average mpg: 28

S4 SPRINT

To all intents and purposes the same as its forebears except for:
Engine:
Carburettors: Twin Weber 40 DCOE 31
Max power (Bhp/rpm): 126/6,500
Max torque (lbs ft/rpm): 113/5,500
Compression ratio: 10.3:1
Performance figures (mph/secs)
0-60: 7.0
0-100: 20.7
1/4-mile (secs): 15.0
Max speed (mph): 118
Average mpg: 25.5

TYPE 28 LOTUS FORD CORTINA

Engine: Lotus-Ford
Cylinders: 4
Fuel supply: Twin Weber 40DCOE carburettors
Bore/stroke (mm): 82.5/72.8
Cubic capacity (cc): 1,558
Compression ratio: 9.5:1
Valve gear: Twin OHC
Ignition: Coil and distributor
Main bearings: 5
Max power (Bhp/rpm): 105/5,500
Max torque (lbs ft/rpm): 108/4,000

Transmission (ratios/mph per 1,000rpm)
        upto July '64, '64 to Oct '65, Post '65
1st: 2.57 3.54 2.97
2nd: 1.64 2.04 2.01
3rd: 1.23 1.41 1.397
4th: 1.0 1.0 1.0
Reverse: 2.807 3.963 3.324

Final drive: 3.9:1
Wheels & Tyres: 5.5Jx13 steel wheels, 165x13-ins
Brakes (F/R): 9.5ins discs/9x1.75ins drums
Steering: Recirculating ball
Lock-to-lock: 3
Turning circle(ft): 36
Measurements (ins)
Length: 168.3
Width: 62.5
Height: 53.4
Wheelbase: 98.4
Track (F/R): 51.6/49.5
Ground clearance: 5
Kerb weight (Lbs): 1,652 – 1,848

Suspension
Front: Independent, MacPherson struts, lower arms,
    anti-roll bar coil springs
Rear: 'A' bracket, twin trailing arms, coil over shock
    absorbers (from July '65, leaf springs and radius
    rods).
Performance figures (mph/secs)
0-60: 9.9
0-100: 33.5
1/4-mile (Secs): 17.4
Max speed (Mph): 106
Average mpg: 21
Production figures: 1,797 (road-going cars only)

TYPE 46, 54, 74 LOTUS EUROPA

Engine: Renault 16
Cylinders: 4 in-line
Fuel supply: Twin-choke Solex 35 DIDSA carburettor
Bore/stroke (mm): 76/81
Cubic capacity: 1470
Compression ratio: 10.25:1
Valve gear: Overhead, pushrod and rockers
Ignition: Coil and distributor
Main bearings: 5
Max power (bhp/rpm): 78/6,000

Max torque (lbs ft/rpm): 76/4,000

Transmission (ratios/mph per 1,000rpm)
1st: 3.61/5.1
2nd: 2.25/8.2
3rd: 1.48/12.5
4th: 1.03/17.9
Reverse: 3.08
Final drive: 3.56:1
Wheels & tyres: Pressed steel, 4.5ins rims. 155-13ins
    tyres.
Brakes (F/R): 9.75ins discs/8x1.5ins drums
Steering: Rack-and-pinion
Lock-to-lock: 2.25
Turning circle (ft): 33.75
Measurements (ins)
Length: 156.5
Width: 64
Height: 42
Wheelbase: 91
Track (F/R): 53/53

Suspension
Front: Independent coil springs, double wishbones,
    dampers.
Rear: Independent radius arms, transverse links, fixed-
    length drive shafts, dampers.

Performance figures (mph/secs)
0-60:   9.5
0-100:   30.3
1/4-mile: 17.3
Max sped (mph):   14.7
Average mpg: 29.8
Production figures: 9230 (all models).
Type 74 Lotus Europa and Europa Special(*)
Engine: Lotus-Ford iron block/alloy head
Cylinders: 4
Fuel supply: Twin Dellorto 40 DHLA
Bore/stroke (mm): 82.5/72.7
Cubic capacity (cc): 1,558
Compression ratio: 9.5:1 10.3:1*
Valve gear: Twin OHC
Ignition: Coil and distributor
Main bearings: 5
Max power (Bhp/rpm): 105/6,000 126/6,500*
Max torque (lbs ft/rpm): 103/4,500 113/5,500*

Transmission (ratio/mph per 1,000rpm)
1st: 3.61/5.2 3.62/4.9*
2nd: 2.26/8.3 2.33/7.6*
3rd: 1.48/12.7 1.60/11.1*
4th: 1.03/18.3 1.21/14.7*
5th: 0.87/20.4*
Reverse: 3.08 3.08*
Final drive ratio: 3.56:1 3.77:1*
Wheels and tyres: 5.5Jx13ins. 175/70-13 (front),
    185/70-13 (rear).
Brakes (F/R): 9.25ins discs/8x1.25ins drums/8x1.5ins
    drums*
Steering: rack-and-pinion
Lock-to-lock: 2.3
Turning circle (ft): 36
Measurements (ins)
Length: 157

Width: 65
Height: 43.25
Wheelbase: 91
Track (F/R): 54.25/54.25
Ground clearance: 4.25
Kerb weight (lbs): 1,557
Performance figures (mph/secs)
0-60:   8.2, 6.6*
0-100:   25.5, 21.6*
1/4-mile: 16.1, 14.9*
Max speed (mph): 120, 123.3*
Average mpg: 24.2, 24.2*

TYPE 50 LOTUS ELAN PLUS 2, PLUS 2S 130(*), PLUS 2S 130/5 (**)

Engine: Lotus-Ford
Cylinders: 4
Fuel supply: Twin Weber 40 DCOE18
Bore/stroke (mm): 82.6/72.8
Cubic capacity (cc): 1,558
Compression ratio: 9.5:1 10.3:1*
Valve gear: Twin OHC
Ignition: Coil and distributor
Main bearings: 5
Max power (Bhp/rpm): 118/6,000 126/6,500*
Max torque (lbs ft/rpm): 108/4,000 113/5,500*

Transmission (ratio/mph per 1,000rpm)
1st: 2.97/ 2.97/6.0* 3.2/5.6**
2nd: 2.01/ 2.01/8.9 2.0/8.9
3rd: 1.40/ 1.40/12.6 .37/13.0
4th: 1.0/ 1.0/17.8 1.0/17.9
5th: 0.8/22.4
Reverse: 3.32 3.32 3.46
Final drive: 3.70:1 3.77:1 3.77:1*
Wheels and tyres: Pressed steel, 5.5ins. 165-13ins
Brakes (F/R): Discs 10ins/10ins
Steering: Rack-and-pinion
Lock-to-lock: 2
Turning circle (ft): 28
Measurements (ins)
Length: 169
Width: 66
Height: 47
Wheelbase: 96
Track (F/R): 54/55
Ground clearance: 6
Kerb weight (lbs): 2,086
    Suspension (F/R): Independent, wishbones, coil
    springs, telescopic dampers.
Performance figures (mph/secs)
0-60: 8.4 7.7*   7.6**
0-100: 24.2   23.0*   24.0**
1/4-mile: 16.4   15.9*   16.0**
Max speed (mph): 122 121* 122**
Average mpg: 28 26.1* 25**
Production figures: Approximately 3,300

TYPE 75 ELITE

Engine: Lotus Type 907
Cylinders: 4
Fuel supply: Twin Dellorto DHLA45 carburettors
Bore/stroke (mm): 95.3/69.2
Capacity (cc): 1,973
Compression ratio: 9.5:1
Valve gear: 4 valves per cylinder, DOHC
Ignition: Coil and distributor
Main bearings: 5
Max power (Bhp/rpm): 155/6,500 (early cars)
    160/6,200
Max torque (lbs ft/rpm): 135/5000 (early cars)
    140/4,900

Transmission (ratios/mph per 1,000rpm)

1st: 3.2/5.7
2nd: 2.02/9.1
3rd: 1.37/13.3
4th: 1.1/18.3
5th: 0.803/22.9
Reverse: 3.460
Final drive: 3.73 (on early cars) 4.1:1 standardised
Wheels and tyres: 7ins alloys, 205/60VR14.
Brakes (F/R): 10.4 ins discs/9x2.25ins drums
Steering: Power-assisted, rack-and-pinion
Turns lock-to-lock: 3.5
Turning circle (ft): 36.25
Measurements (ins)
Length: 175.5
Width: 71.5
Height: 47.5
Wheelbase: 97.8
Track (F/R): 58.5/59
Ground clearance: 4.6
Kerb weight (lbs): 2,440

Suspension
Front: Independent, coil springs, upper wishbone with
    single lower link and anti-roll bar, telescopic
    dampers
Rear: Independent, tapered coil springs, bottom
    wishbone formed by diagonal trailing arm and
    lateral link, fixed drive shaft, telescopic dampers
Performance figures (mph/secs)
0-60:   7.8
0-100:   24.5
1/4-mile: 16.4
Max speed (mph): 126
Average mpg: 20.9

Production figures: 2,530
Types 502 and 503 mechanically identical to original
    Type 501. Type 504 differed only in having Borg-
    Warner 65 three-speed automatic.
Transmission (ratios/mph per 1,000rpm)
1st: 2.39:1/7.7
2nd: 1.45:1/12.6
Top: 1.00:1/18.3
Reverse: 2.09:1
Final drive ratio: 3.73:1
Performance figures (mph/secs)
0-60: 10.4

0-100mph: 31.6
1/4-mile: 17.3
Max speed (mph): 120
Average mpg: 19.1
S2.2, same overall specifications as for the '500' series
    except for:
Engine: Lotus Type 912
Cylinders: 4
Fuel supply: Twin Dellorto DHLA 45E carburettors
Bore/stroke (mm): 95.3/76.2
Capacity (cc): 2174
Valve gear: 4 valves per cylinder, DOHC
Ignition: Breakerless
Main bearings: 5

Transmission (ratio/mph per 1,000rpm)
1st: 2.96/5.6
2nd: 1.93/8.62
3rd: 1.39/11.97
4th: 1.0/16.64
5th: 0.813/20.47
Reverse: 3.705:1
Final drive: 4.1

Performance figures (mph/secs)
0-60: 7.5
0-100: 22.5
1/4-mile: 16.1
Max speed (mph): 129
Average mpg: 20.6

TYPE 76 LOTUS ECLAT

Mechanically identical to the Elite bar the following
    exceptions:
Type 520 and Eclat Sprint* Getrag 5-spd**
Transmission (ratios/mph per 1,000rpm)

1st:	3.16/5.6	3.20/5.7*	2.96/6.2**
2nd:	1.94/9.1	2.00/9.1	1.93/8.6
3rd:	1.41/12.6	1.37/13.3	1.39/13.1
4th:	1.00/17.9	1.00/18.3	1.00/18.3
5th:		0.80/22.9	0.81/22.4
Reverse:	3.35	3.47	3.71
Final drive:		3.73:1	3.73:1

Wheels and tyres: 5.5ins pressed steel rims,
    185/70HR-13in tyres 5.5ins alloy rims*
Performance figures (mph/secs)
0-60: N.A.
0-100: N.A.
1/4-mile: N.A.
Max speed (mph): N.A.
Average mpg: N.A.

TYPE 79, LOTUS ESPRIT

Engine: Lotus Type 907
Cylinders: 4
Fuel supply: Twin Dellorto DHLA45 carburettors
Bore/stroke (mm): 95.3/69.2
Capacity (cc): 1,973

Compression ratio: 9.5:1
Valve gear: 4 valves per cylinder, DOHC
Ignition: Coil and distributor
Main bearings: 5
Max power (Bhp/rpm): 160/6.200
Max torque (lbs ft/rpm): 140/4,900

Transmission (ratio/mph per 1,000rpm)
1st: 2.92:1/5.7
2nd: 1.94:1/8.56
3rd: 1.32:1/12.58
4th: 0.97:1/17.15
5th: 0.76:1/21.85
Reverse: 3.46
Final drive: 4.35:1

Wheels and tyres:
F/R: 205-60HR14, 6ins rims/205/70HR14, 7ins rims
Brakes (F/R): 9.7ins discs/10.6ins discs
Steering: Rack-and-pinion
Lock-to-lock: 3.1
Turning circle (ft):
Measurements (ins)
Length: 165
Width: 73.25
Height: 43.7
Wheelbase: 96
Track (F/R): 59.5/59.5
Ground clearance: 5.5
Kerb weight (lbs): 2,218
Suspension
Front: Independent, coil springs, wishbones, anti-roll
    bar, telescopic dampers
Rear: Independent, coil springs, lower
    wishbones/semi-trailing arms, fixed length
    driveshafts, telescopic dampers
Performance figures (mph/secs)
0-60: 8.4
0-100: 27.4
1/4-mile (secs): 16.3
Average mpg: 23.3
Production figures: 2,919 (normally-aspirated cars
    only)
Esprit S2
Same mechanical specification to the S1 except for:
    7ins front
wheels. 7.5ins rear wheels. Compact spare wheel,
    185/70HR-13 on 5.5ins
wheel. Kerb weight: 2334 lbs.
Esprit S2.2
Same mechanical specification to the S2 except for:
Engine: Lotus Type 912
Cylinders: 4
Fuel supply: Twin Dellorto DHLA 45E carburettors
Bore/stroke (mm): 95.3/76.2
Capacity (cc): 2174
Compression ratio: 9.4:1
Valve gear: 4 valves per cylinder, DOHC
Ignition: Breakerless
Main bearings: 5
Max power (Bhp/rpm): 160/6,500
Max torque (lbs ft/rpm): 160/5,000

Transmission (ratio/mph per 1,000rpm
1st: 2.92:1/5.7

2nd: 1.94:1/8.56
3rd: 1.32:1/12.58
4th: 0.97:1/17.15
5th: 0.76:1/21.85
Reverse: 3.46
Final drive: 4.35:1
Performance figures (mph/secs)
0-60: N.A.
0-100: N.A.
1/4-mile (secs): N.A.
Max speed (mph): N.A.
Average mpg: N.A.

TYPE 82 LOTUS TURBO ESPRIT

Same mechanical specification to the S2.2 except for:
Engine: Lotus Type 910
Cylinders: 4
Fuel supply: Twin Dellorto 40DHLA carburettors and
    Garrett AiResearch T3 turbocharger
Max boost (psi): 8
Bore/stroke (mm): 95.3/76.2
Capacity (cc): 2174
Compression ratio: 7.5:1
Valve gear: 4 valves per cylinder, DOHC
Ignition: Breakerless
Main bearings: 5
Max power (Bhp/rpm): 210/6,000
Max torque (lbs ft/rpm): 200/4,000
Transmission (ratio/mph per 1,000rpm)
1st: 2.92/5.69, 5.88*
2nd: 1.94/8.56, 8.84*
3rd: 1.32/12.58, 13.00*
4th: 0.97/17.15, 17.69*
5th: 0.76/21.85, 22.58*
Reverse: 3.46
Final drive: 4.37:1
Wheels and tyres: Larger 15ins wheels* (optional on
    S3), standard
Performance figures (mph/secs)
0-60:   6.1
0-100:   17.0
1/4-mile (secs): 14.6
Max speed (mph): 148
Average mpg: 18.0

TYPE 85, LOTUS ESPRIT S3

Same mechanical specification to the S2.2 except for:
Suspension
Rear: Coil springs,lower wishbones/semi-trailing arms,
    upper links and telescopic dampers
Brakes (F/R): 10.5ins discs/10.8ins discs
F/R: 195/60VR-15ins on 7ins alloys/235/60VR-15ins on
    8ins alloys.
Spare tyre, 175/70SR-14ins
Measurements (ins)
Track (F/R) with optional wheel/tyres: 60.5/61.2
Kerb weight (lbs): 2,352
Performance figures (mph/secs)
0-60: 6.7

0-100: 20.9
1/4-mile (secs): 15.5
Max speed (mph): 134
Average mpg: 21

X180 – LOTUS ESPRIT

Peter Stevens' designed Esprit
Engine:
Cylinders: 4
Fuel supply: Two twin-choke Dellorto carburettors
Bore/stroke (mm): 95.28/76.2
Cubic capacity: 2174
Compression ratio: 10.9:1
Valve gear: 4 valves per cylinder, DOHC
Ignition: Breakerless
Main bearings: 5
Max power (bhp/rpm): 172/6,500
Max torque (lbs ft/rpm): 163/5,000
Transmission (ratio/mph per 1,000rpm)
1st: 3.36/5.8
2nd: 2.05/9.5
3rd: 1.38/14.1
4th: 1.03/18.7
5th: 0.82/23.7
Reverse: 3.54
Final drive ratio: 3.88:1
Wheels and tyres (F/R): 195/60VR-15ins, 7ins cast
    alloys/235/60VR-15ins, 8ins cast alloys
Brakes (F/R): 10.1 ins/10.8ins discs
Steering: Rack-and-pinion
Lock-to-lock: 3.0
Turning circle (ft): 34.75
Measurements (ins)
Length: 170.5
Width: 73.2
Height: 44.8
Wheelbase: 96.7
Track (F/R): 60/61.2
Ground clearance:
Kerb weight (lbs): 2,590
Suspension
Front: Independent, coil springs, wishbones, anti-roll
    bar, telescopic dampers
Rear: Independent, coil springs, lower
    wishbones/semi-trailing arms, upper links,
    telescopic dampers
Performance figures (mph/secs)
0-60: 6.5
0-100: N.A.
1/4-mile (secs): N.A.
Max speed (mph): 138
Average mpg: 30.8
Production figures: 3,012 (+1993's figures)

LOTUS ESPRIT TURBO

Mechanically identical to the normally-aspirated X180
    except for:
Engine
Fuel supply: Two twin Dellorto carburettors, Garrett
    AiResearch turbocharger
Boost pressure (psi): 9.5
Compression ratio: 8:1
Max power (bhp/rpm): 215/6,500
Max torque (lbs ft/rpm): 220/4,250

Transmission (ratio/mph per 1,000rpm)
1st: 3.36/5.8
2nd: 2.06/9.5
3rd: 1.38/14.2
4th: 1.04/18.9
5th: 0.82/23.9
Reverse: 3.54
Final drive ratio: 3.89
Kerb weight (lbs): 2,800
Performance figures (mph/secs)
0-60: 5.4
0-100: 13.3
1/4-mile (secs): 13.7
Max speed (mph): 150
Average mpg: 19.6

ESPRIT TURBO SE

Mechanically identical to the Turbo except for:
Engine: Lotus Type 910S
Fuel supply: Electronic MPFI, Garrett AiResearch
    TB03 turbocharger water-cooled with integral
    wastegate and chargecooler
Boost pressure (bar): 0.85
Compression ratio: 8.0:1
Ignition: Electronic distributorless twin-coil
Max power (bhp/rpm): 264/6,500
Max torque (lbs ft/rpm): 261/3,900
Transmission (ratio/mph per 1,000 rpm)
1st: 3.36:1/5.6
2nd: 2.05:1/9.2
3rd: 1.38:1/13.7
4th: 1.03:1/18.4
5th: 0.82:1/23.1
Reverse:
Final drive ratio: 3.889:1
Wheels and Tyres (F/R): 7Jx15ins alloys, 215/50ZR-15
    / 5Jx16ins alloys, 245/50ZR-16
Brakes (F/R): 10.2ins ventilated discs/10.8ins discs
Steering: Rack-and-pinion
Lock-to-lock: 3.0
Turning circle (ft): 36
Performance figures (mph/secs)
0-60: 4.7
0-100: 11.9
1/4-mile (secs): 13.5
Max speed (mph): 163
Average mpg: 23.5

ESPRIT S

Same mechanical specifications as for the SE, bar:
Engine
Fuel supply: Delco MPFI, Garrett AiResearch TBO3
 turbocharger
Boost pressure (bar): 0.82
Compression ratio: 8.0:1
Ignition: GM ELectronic Multi-point
Max power (bhp/rpm): 228/6,500
Max torque (lbs ft/rpm): 218/4,000

Transmission (ratio/mph per 1,000rpm)
1st: 3.36/5.6
2nd: 2.06/9.2
3rd: 1.38/13.7
4th: 1.04/18.2
5th: 0.82/23.1
Reverse: 5.4
Final drive ratio: 3.89:1
Performance figures (mph/secs)
0-60: 5.2
0-100: 12.7
1/4-mile (secs): N.A.
Max speed (mph): 155
Average mpg: N.A.
Lotus Esprit
Specification as per original X180 Esprit Turbo, but
 glassback deleted.

LOTUS ESPRIT S4

Mechanically unchanged from predecessor with the
 exception of revised suspension geometry, ABS
 braking and power-assisted steering. Bodywork
 and interior also revised. New alloy 17ins rims
 with 215/40ZR 17x7J front tyres and 245/45ZR
 17x8.5J rear tyres.
Measurements (ins)
Length: 172
Width: 73.5
Height: 45.2
Wheelbase: 95.9
Track (F/R): 60.1/60.8
Lotus Sport 300
Mechanically the same as the Esprit S4 with the
 following exceptions:
Engine
Max power (bhp/rpm): 302/6,400
Max torque (lbs ft/rpm): 287/4,400
Transmission (ratio/mph per 1,000rpm)
1st: 3.36/5.6
2nd: 2.05/9.2
3rd: 1.38/13.7
4th: 1.03/18.4
5th: 0.82/23.1
Reverse:
Final drive ratio: 3.89:1
Wheels and Tyres (F/R): Alloy 8.5x16, 245/45ZR16 /
 Alloy 10.5x17, 315/35ZR17
Brakes (F/R): 13ins/11ins ventilated discs, and Delco
 Moraine anti-lock system
Kerb weight (lbs): 2,738

Performance figures (mph/secs)
0-60: 4.7
0-100: 11.7
1/4-mile (secs): 13.4
Max speed (mph): 161
Average mpg: 22

TYPE 89 LOTUS EXCEL

Mechanically identical to the Elite/Eclat 2.2 bar the
 following exceptions:
Transmission (ratios/mph per 1,000rpm)
1st: 3.285/5.2
2nd: 1.894/9.0
3rd: 1.275/13.3
4th: 1.000/17.0
5th: 0.783/21.7
Reverse: 4.5
Final drive ratio: 4.1:1
Wheels & tyres: 7ins cast alloy rims, 205/60VR-14ins
 tyres
Brakes (F/R): 10.16ins/10.47ins discs
Steering: Rack-and-pinion, power-assistance optional
Measurements (ins)
Length: 172.3
Track (F/R): 57.5/57.5
Kerb weight (lbs): 2,503
Suspension
Front: Independent, coil springs, wishbones, anti-roll
 bar, telescopic dampers
Rear: Independent, coil springs, lower wishbones,
 upper links, telescopic dampers
Performance figures (mph/secs)
0-60mph: 7.1
0-100mph: 20.1
1/4-mile(secs): 15.4
Max speed (mph): 130
Average mpg: 19.5
Production figures: 1,327 manufactured
Excel SE
Same mechanical specifications as for the Excel bar:
Engine
Compression ratio: 10.9:1
Max power (bhp/rpm): 180/6,500
Max torque (lbs ft/rpm): 165/5,000
Transmission (ratio/ mph per 1,000 rpm)
1st: 3.285/5.01
2nd: 1.894/8.69
3rd: 1.275/12.91
4th: 1.000/16.46
5th: 0.783/21.02
Reverse: 4.4
Final drive: 4.1:1
Performance figures (mph/secs)
0-60: 6.8
0-100: 19.9
1/4-mile (secs): 15.3
Max speed (mph): 134
Average mpg: 19.6

EXCEL SA

Same mechanical specification as the SE except for:
Transmission (4-speed ZF HP22 automatic)
1st: 2.73/6.7
2nd: 1.56/11.7
3rd: 1.00/18.2
Top: 0.73/25.0
Reverse: 7.79
Final drive ratio: 3.727:1
Top gear: 24.9mph/1,000rpm
Performance figures (mph/secs)
0-60: 8.2
0-100: 24.9
1/4-mile (secs): 16.9
Max speed (mph): 124
Average mpg: 22

M100 –LOTUS ELAN SE & LOTUS ELAN(*)

Engine: Isuzu Lotus
Cylinders: 4
Fuel supply: Electronic MPFI with IHI water-cooled
 turbocharger and air-to-air intercooler. Electronic
 MPFI*
Boost pressure (psi): 9.4
Bore/stroke(mm): 80/79
Cubic capacity: 1,588
Compression ratio: 8.2:1
Valve gear: Four valves per cylinder, DOHC
Ignition: Electronic distributorless
Max power (bhp/rpm): 165/6,600 130/7,200*
Max torque (lbs ft/rpm): 148/4,200 105/4,200*
Transmission (ratio/mph per 1,000rpm)
1st: 3.33/5.2 3.33/4.8*
2nd: 1.92/9.0 1.92/8.4*
3rd: 1.33/12.99 1.33/12.0*
4th: 1.03/16.9 1.03/15.7*
5th: 0.83/20.9 0.83/19.5*
Reverse: 3.583
Final drive ratio: 3.833:1 4.117:1*
Wheels & Tyres: Alloy, 6.5Jx15. 205/50ZR15
Brakes (F/R): Ventilated 10ins discs/solid 7.3ins discs
Steering: Power-assisted (SE only) rack-and pinion
Lock-to-lock: 2.9 3.1*
Turning circle (ft): 35
Measurements (ins)
Length: 149.7
Width: 74.3
Height: 48.4 (hood raised)
Wheelbase: 88.6
Track (F/R): 58.5/58.5
Ground clearance: 5.1
Kerb weight (lbs): 2249 2198*
Suspension
Front: Independent by unequal length wishbones,coil
 springs and dampers. Anti-roll bar. Aluminium alloy
 subframes.
Rear: Independent by upper link and wide-based
 lower wishbone.
Springs and dampers, anti-roll bar.
Performance figures (mph/secs)
0-60: 6.7 7.6

0-100: 17.5 xx.x
1/4-mile (secs): 15.0 16.1
Max speed (mph): 137 122
Average mpg: 33.4 33.9
Production figures: 385

M104 – LOTUS CARLTON/OMEGA

Engine: Lotus-GM
Cylinders: 6
Fuel supply: Electronic MPFI, twin Garrett AiResearch
    T25 turbochargers and chargecooler
Bore/stroke(mm): 95/85
Capacity (cc): 3,615
Compression ratio: 8.2:1
Valve gear: 4 valves per cylinder, DOHC
Ignition: Distributorless electronic with three coils
Max power (bhp/rpm): 377/5,200
Max torque (lbs ft/rpm): 419/4,200
Transmission (ratio/mph per 1,000rpm)
1st: 2.68/8.2
2nd: 1.80/12.2
3rd: 1.29/17.0
4th: 1.00/22.0
5th: 0.75/29.4
6th: 0.50/44.1
Reverse:
Final drive ratio: 3.45:1
Wheels & Tyres (F/R): Alloy, 17x8.5ins 235/45ZR17 /
    Alloy, 17x9.5ins 265/40ZR17
Brakes(F/R): 12.9/11.8ins ventilated discs. ABS
Steering: Recirculating ball, power-assisted.
Lock-to-lock: 2.6
Turning circle (ft): 32.6
Measurements (ins)
Length: 187.7
Width: 76.1
Height: 56.5
Wheelbase: 107.5
Track (F/R): 57/58
Ground clearance: 7
Kerb weight (lbs): 3,641
Suspension
Front: Independent, MacPherson struts, coil springs,
    twin-tube dampers, anti-roll bar
Rear: Multi-link, semi-trailing arms, progressive rate
    coil springs, twin-tube dampers, self-levelling
Performance figures (mph/secs)
0-60: 5.1
0-100: 11.1
1/4-mile (secs): 13.5
Max speed (mph): 176
Average mpg: 22.3
Production figures: 950

Chapman stayed true to his principles and retained the backbone chassis concept that had been the basis of the original Elan and the Europa; this is the third Elan Plus 2 chassis. (Monitor)